About Island Press

Since 1984, the nonprofit organization Island Press has been stimulating, shaping, and communicating ideas that are essential for solving environmental problems worldwide. With more than 1,000 titles in print and some 30 new releases each year, we are the nation's leading publisher on environmental issues. We identify innovative thinkers and emerging trends in the environmental field. We work with world-renowned experts and authors to develop cross-disciplinary solutions to environmental challenges.

Island Press designs and executes educational campaigns, in conjunction with our authors, to communicate their critical messages in print, in person, and online using the latest technologies, innovative programs, and the media. Our goal is to reach targeted audiences—scientists, policy makers, environmental advocates, urban planners, the media, and concerned citizens—with information that can be used to create the framework for long-term ecological health and human well-being.

Island Press gratefully acknowledges major support from The Bobolink Foundation, The Curtis and Edith Munson Foundation, The Forrest C. and Frances H. Lattner Foundation, The Freedom Together Foundation, The Kresge Foundation, The Summit Charitable Foundation, Inc., and many other generous organizations and individuals.

The opinions expressed in this book are those of the author(s) and do not necessarily reflect the views of our supporters.

LIVING OFF-GRID

LIVING OFF-GRID

50 Steps to Unplug, Become Self-Sufficient, and Build the Homestead of Your Dreams

RYAN MITCHELL

ISLANDPRESS | Washington | Covelo

Library of Congress Control Number: 2025934646

All Island Press books are printed on environmentally responsible materials.

Manufactured in the United States of America
10 9 8 7 6 5 4 3 2 1

Keywords: Black water; Cooking off-grid; Food preservation; Food security; Gray water; Heating and cooling your home; Home building; Home gardening; Homesteading; Purchasing land; Self-sufficiency; Solar panels; Sourcing water; Sustainable design; Tiny house; Well digging; Wind power

To Dad. Your love and wisdom continue to guide me through my journey.

CONTENTS

—

INTRODUCTION WHY DISCONNECT?

—

It was New Year's Eve, and after two weeks of overcast days, my batteries were finally tapped out. The power went off in my house, and that led me to pull out my backup generator, which I had for this exact purpose, but it didn't want to start.

I was working on the generator at 11:00 at night; it was 35 degrees out and the heaviest downpour of the year. My hands stung as I broke down the carburetor. All the while, I was trying to keep water out of it while I cleared the clog in the jet. I was fiddling with all the little pieces and trying not to lose one in the dark, but my fingers barely worked after being too cold for too long.

It was miserable.

In that moment—and many others, if I'm being honest—I cursed my decision to live off the grid.

But as I put the carburetor back together, gave it a little spray of starting fluid, and ripped the cord, the generator roared to life. The lights went back on in my house, and the batteries of my solar setup began to charge. I stomped back inside, soaked to the bone, for a hot shower.

So, is living off the grid worth it? If you asked me on that particular night, maybe not. If you asked me today, after not having had a power bill in over a decade, definitely.

I like telling this story to help others realize that unplugging from society's standard systems—public utilities, sewage systems, housing developments—and instead creating your own self-sufficient homestead isn't always the fantasy you imagined. But if you plan ahead and set yourself up for success, you'll have more good days than bad.

Despite all the hurdles, I now live on 11 acres of land, in a house that I designed and that I power with solar; I get my water from a well, have a composting toilet and septic system, eat from my garden, raise baby quail, and run my own business. My expenses are a fraction of what they were when I paid rent and utilities, and I have infinitely more free time to pursue the things that fulfill me.

I never thought I'd live this way. Like many, I had followed the common path, going to school, getting good grades, heading off to college, and then later to graduate school. I landed my first adult job, believing that with a degree and hard work, everything would fall into place.

This is, after all, the American dream: pay your dues and move up the ranks, earn a good salary, and ultimately receive your reward: a nice home with a white picket fence, married with 2.5 kids, and a red convertible parked in the driveway that you jokingly refer to as your midlife crisis.

That was my worldview right up until, six months into my career, the Great Recession hit. The company I worked for closed its doors one Friday afternoon, leaving everyone standing in the parking lot with stunned looks on our faces while holding cardboard boxes filled with the contents of our desks.

It was there, in that parking lot, that I promised myself I'd never find myself in that position again. I realized that I had been following a script—a script that I had not given much thought to, but instead blindly followed because it's what you "should" do.

I won't disparage that lifestyle out of hand because, for some, it works just fine. But I will say that it's essential for you to be the decider in your own life. You want to be intentional about your choices: where you live, how you spend your time and money, how you interact with the broader world. If you don't, others will be happy to make those decisions for you, to their benefit.

This doesn't need to be some boisterous protest against the status quo, but a quiet consideration of what is important to you. For me, simply deciding what I wanted for myself was the most impactful thing I've ever done. After leaving that parking lot with my cardboard box, I set about building my new life.

That new life led me to build a house nestled in the woods, away from the hustle and bustle of the city. I had decided that by building my own home, I could skirt the need for a mortgage entirely. I chose the location because, after a year of reflection, I discovered I felt most comfortable in the quiet of nature.

It was here that I realized that pursuing a simpler life was often filled with complexities.

In my case, I wanted to live in the countryside, away from it all. But that came with its own set of challenges. I found out that city hall wouldn't grant me a permit for a septic system, but instead wanted me to connect to the sewer lines. The first quote I got was for tens of thousands of dollars and didn't include the government's fee of $11,000 for the privilege of installing the meter to facilitate their charging me monthly for utilities.

I also learned that the power company wanted tens of thousands of dollars to run the power lines to my house. All these were significant setbacks because I didn't have that much money to my name.

It was then that I went searching for solutions and discovered a new way to solve these needs: going off the grid.

Your own interest in disconnecting may be different from mine, but you're likely here because you also want something better for yourself. In this book, I speak about things like self-sufficiency, empowerment, ownership, intentional living, and a slower way of being. There are many reasons that you might be drawn to living off-grid, from saving money and learning new skills to living closer to nature and shrinking your environmental footprint. Whatever your motivation, make no mistake, this is a lifestyle. What exactly that means for you is something I'll help you discover for yourself, then give practical advice for each part of your journey.

This book is structured in three main sections. The first section discusses adopting the mindset you'll need to pursue an off-grid lifestyle. The second section will help you develop a vision of your goals and then set up a plan to make them happen, including buying land. Finally, the last section gives my advice on approaching many of the key aspects of the life you want to build, including heating and cooling your home, handling gray and black water, doing laundry off the grid, and even aging in place.

It's important to realize that pursuing this lifestyle isn't something that will happen overnight, and it's not without its challenges, but if you're here reading this book, something tells me you're not one to shy away from hard work.

This book is for anyone flirting with the idea of living off the grid or those who have already gone down the proverbial rabbit hole. It is designed to be a wayfinding tool that will help you determine a direction and gives you the key milestones on the way to your final destination, from simply saving some money and carbon with a few small solar arrays to developing a completely independent homestead.

I'll also advise those who are in a variety of scenarios. Whether you live in a small apartment in a city, a small lot in suburbia, or on acreage in the countryside, you'll find this book useful all the same.

I also want to point out that while this book focuses on off-grid living, the overlap with homesteading is substantial; I'll broadly speak to both and use the two terms interchangeably. While off-grid living and homesteading aren't quite the same, I find the differences to be more a matter of semantics, without much practical distinction.

Those who are interested in one are usually interested in the other. The one main difference might be that of food production, mainly through gardening and livestock. Off-grid living focuses on developing a self-sufficient life, and eating food is the core of living, which is why I think that producing both your own power and your own food are equally important.

The nature of this information means that we'll have to keep the discussion at a relatively high level. Although entire volumes could be written about any one of the topics I'll touch on, my goal isn't to provide you with a precise prescription but rather with the knowledge you'll need to forge your own path. After all, if you wanted to follow someone else's rules or take a cookie-cutter approach, you probably wouldn't consider living off the grid anyway. But while unplugging necessarily leads you down a road less traveled, you're not alone in this journey.

PART I

ADOPT THE OFF-GRID MINDSET

STEP 01   REALIZE OFF-GRID BENEFITS

—

We wouldn't choose the more difficult path unless we thought it was worth it. While we each have our own reasons for wanting this lifestyle, people often cite a few common motivations for disconnecting from the grid.

Going against the grain is not without its challenges, but I've learned that with enough resolve and a little bit of grit, this is all achievable, and the payoff is enormous. When I finally moved into my off-grid house, I went from paying $1,500 a month in rent and utilities to $15 per month. And yes, that is not a typo.

Having flipped the script on life, I saw many advantages you can enjoy, too. I'll go into more depth on each of these, but here are some of the main benefits of going off the grid:

- Simplifying your life
- Going green
- Developing self-reliance
- Escaping debt
- Eating healthfully
- Living on your own terms
- Building community.

STEP 02   SIMPLIFY YOUR LIFE

—

One of the main reasons that people decide to live off the grid is that it's an antidote to the chaos of the modern-day rat race. Many of us are glued to our devices, the world is moving faster and faster, and people brag about being busy like it's a badge of honor. All this hasn't worked out so well for us.

People are finding their quality of life going in the wrong direction, and their happiness with it. We're plagued by debt, unhealthy food is poisoning our bodies, and society is seeing an increasing level of mental health crises. It can be unsettling when we take stock of the state of our world.

Despite these troubling trends, the fact that you're here, reading this book, tells me that you're an optimist. After all, you don't sign up for years of hard work to create your own homestead if you think it will all be for naught. That spirit of building for the future is summed up nicely in the adage: "A society grows great when old men plant trees in whose shade they shall never sit."

Even if you're a cynic, the best practical course of action is to work to build a better life for yourself and those you love—regardless of what is going on in the wider world. As individuals, we may not have the power to change governments or institutions on our own, but we can take concrete steps to improve our situations and those of the people around us. The cosmic joke for a true skeptic is that positive action is the best bottom-up approach I've encountered. If enough people take ownership and lead through example, change can filter up.

Taking ownership means slowing down and being intentional. In fact, intentionality is the great superpower of modern life. In an age where we are equipped with endless information at our fingertips, if we just pause long enough to figure out what is right for us, we can leverage that information in the best way possible.

For many, myself included, being intentional leads to a simpler way of life. The idea of simplifying can have many meanings to different folks, but for me, it's about focusing on the things that are important and reducing the things that are not as important. Figure out what you want more of and do that; figure out what you want less of and don't do that. It's simple.

Figuring out what is and isn't important to you is challenging when you are too busy to think for yourself. I find that it takes at least a few hours of slowness to get my mind to settle before doing the deep work that is uncovering what matters to me. You might be different, but we'll go through how to do this in a later part of this book.

Living a simpler life takes some adjusting to, especially if those around you are still stuck on the hamster wheel. After moving into my off-grid house, my bills were cut by more than half, which meant I didn't have to work as much. I was saving more money than ever, but also working less at the same time.

It left me with a lot of time to fill and the challenge of doing it in a way that was right for me. Around this time, I made the jump to working for myself and I could do it from anywhere; this was before working from home was commonplace. So, I found myself reading books in a cozy coffee shop, taking afternoon walks every day, and grocery shopping at 3:00 p.m., when the store was less crowded.

It was a stark contrast to the way I had lived—and the way all my friends continued to live, working 40, 50, or even 60 hours a week. I found myself chatting with retirees I'd cross paths with on my walks, taking monthlong solo trips at times of year when most had to be at work or school, and I got lost in the books I read.

I had the flexibility to build and tend my gardens instead of struggling to keep the weeds at bay. I could check on my chickens and quail every morning instead of having to run off to a cubical. I spent more time outdoors and in nature, which left me feeling happier. I don't think it was just a placebo effect: studies have found that digging in the soil exposes gardeners to bacteria, which helps with serotonin levels, buffers against stress, and creates other positive health impacts.

In short, all those things I never had time to do, I suddenly had time to do. I sat down and wrote my first published book, setting me on a journey I couldn't have imagined, this book being my eighth. I spent more time with those closest to me, something I barely got to do before this shift because I had to spend hours away on business.

It also gave me a lot of time to think and, more importantly, time to slow down enough to do it with a clear head. I woke in the mornings with energy. I suddenly could make a cup of coffee and enjoy it on my front porch before heading off for the day. I went on long hikes that left me feeling centered and having a better sense of equilibrium. All this was possible as a result of my decision to go off-grid, escape the rat race, and simplify.

STEP 03   GO GREEN

—

Living off the grid often involves things like solar panels, composting toilets, and food you grow yourself. Going green may be a major priority for you or simply a side benefit, but either way, I found that I could live a better life and be more sustainable in one fell swoop.

The biggest step to getting off the grid for me was disconnecting from the power grid. I did this because the cost of getting a power line run to my home was cost-prohibitive, but it also meant that I lessened my impact on the environment.

At the time, most of the power in my area was generated by burning coal. That meant that by cutting the cord, six tons of coal per year didn't need to be burned. It was nice to have the convenience of power without all the impact that normally went along with it. I also found that my solar setup was much more reliable than my area's grid power.

Water was also another big one for me. Living the way I did, I reduced my wastewater by around 3,000 gallons per year. Most of my wastewater was gray water, meaning it wasn't clean enough to drink, but it didn't have any pathogens like those found in sewage. That allowed me to water my landscaping with the gray water, keeping it on my property and not adding to an overburdened water system.

It took a little time to figure out the right mix of replacement soaps that were biodegradable, but eventually, I found some that worked just as well as "regular" soaps, detergents, and shampoos but didn't harm the Earth as they made their way to my landscaping beds.

Living in the woods meant that I didn't have a lot of open space to grow a garden, but I kept a small one at the house and then got involved with a community garden to grow some of my food. The time spent in the dirt was a great way to slow down, and it yielded a lot of veggies that I didn't need to buy at the store.

Bringing your food production home is the ultimate form of local shopping, even if you only cover some of your needs. Growing a garden keeps all those vegetables from being transported across vast distances; plus, you select the inputs so you know exactly what is in your food, and you have the option to grow organically.

Similarly, if you decide to raise livestock, you can do it in a way that aligns with your values. Since you're operating on a smaller scale than commercial farms, you can make choices that would not be practical for large-scale agribusiness—or choices they wouldn't be willing to make. These decisions include feeding your livestock organic grains, providing more space for your animals, and allowing grazing or free-ranging. You'll also be able to ensure the animals' quality of life in a way a large operation could never hope to achieve.

Whether you're tending livestock or installing solar panels, the point is that you can choose approaches that sustain yourself and the planet, rather than simply plugging into systems that may or may not benefit either.

STEP 04   DEVELOP SELF-RELIANCE

—

Becoming more self-reliant is a potent activator in your life. For the most part, we can count on ourselves to have our own best interests at heart. But more than that, developing independence gives you greater control over your circumstances.

Control is an essential part of finding fulfillment. And fulfillment is, to me, more important than even happiness, because happiness is subject to the conditions you find yourself in; being fulfilled has an enduring quality that can carry you through dark times.

When you look at disenfranchised employees, unmotivated teams, or a disgruntled populace, they all have one thing in common: a deep dissatisfaction with their ability to control outcomes. The opportunity to work hard and have it add up to an improvement in one's circumstances is essential. Yet in modern society, which has focused on specialization and segmenting the workforce into silos of people with deep expertise in very narrow fields, it can be difficult to exert broad control.

Going off the grid is opting to be a generalist in all your needs. We can still leverage specialist knowledge through books, YouTube videos, etc., but ultimately, adopting this life means skilling up in many practical ways. That doesn't mean you'll develop all these skills overnight. Being scrappy and figuring things out as you go along is important when you're under the gun and must make do with what you have.

When I started my journey, I had no business doing everything I undertook. I was a white-collar desk jockey who decided to build a house despite not having woodworking experience. I lived in the city when I broke ground on my first garden bed, despite not knowing the first thing about gardening.

It was hard and it was uncomfortable, but soon I found my stride after falling flat on my face so many times. Eventually, I learned enough to become confident that I could figure most problems out on my own. I slept much easier knowing that my successes weren't just due to good fortune or the work of others: I had made them a reality. You can do the same; it's a matter of taking the first step, working through the issues that crop up, and adapting, until relying on your own abilities becomes second nature.

The truth is, when you go off-grid, things will go wrong, stuff will break, and your power will click off at some point. I can't tell you how many times I've checked my bank account and given myself a pat on my back, only to have Murphy come in and take me down a notch or two. Homesteaders might call this grit; farmers would just call it hard work. The point is that the more resilient you are, the better off you will be in life and this journey toward self-reliance.

The best way I've found to become more adaptable is to list the ways that I rely on outside elements (power companies, utilities, grocery stores, banks, etc.) and then look for strategies to offset those. You'll have many more things listed than you could realistically do in a lifetime, but choose a few that you can take practical steps on today, and start slowly.

Of course, you are not an island, and going through life as a lone wolf will only get you so far. There will be parts of your life that are simply out of your control; dwelling on them won't change those facts. But in the areas that are available to you, becoming more self-reliant will bear fruit. You will acquire tools, relationships, and wisdom, and through this accumulation of skills and resources, you will build resiliency.

Another key strategy is to seek out new experiences and challenges actively. Stepping outside of your comfort zone not only broadens your perspective, but also hones your ability to adapt to unfamiliar situations. This could mean taking on different types of projects at work, traveling to new places, and, of course, learning new things on your homestead. Each new experience is an opportunity to develop skills and insights that can make you more adaptable in other areas of your life.

STEP 05   ESCAPE DEBT

—

A song I love includes this lyric: "There is no dollar sign on peace of mind, this I've come to know." It resonated so strongly with me when I first heard it. Debt is a dead weight that hangs heavy on our lives. For most, it's necessary to make larger purchases like a home or land. For those who can work to get free of it, the impact is profound.

Moving into a tiny home let me flip the script on finances and allowed me to start paying down my student loans aggressively, save up to buy a large solar-panel system, and pay for my next vehicle in cash so I didn't have to make car payments anymore.

Escaping those debts gave me the freedom to look forward, which proved empowering. I could save more, invest for retirement, roll with the punches, and take a load of stress off my plate. When tough times did come, I didn't have a pile of bills to pay, but I did have cash in the bank for leverage if needed.

Being debt-free also gives you options that those in debt don't have. There have been times when I have been an employee, and because of my financial security, I've had far more control over my work life. The employer–employee dynamic takes a different form when you don't need their money. Suddenly, bad bosses behave better, business owners treat you with respect, and they value your contributions a lot more.

The decision to focus on financial freedom had far-reaching effects beyond just my bank account. It allowed me to pursue passions and hobbies, like travel and woodworking, that I had previously put on the back burner due to financial constraints. I had more time and less stress, leading to a healthier and more fulfilling life.

It allowed me to spend more time with the people that matter most to me. My relationships improved as I was no longer constantly preoccupied with working extra hours to offset financial worries or working to pay the bills, enabling me to be more present and engaged with my loved ones.

Additionally, the journey toward debt-free living taught me valuable lessons in discipline and patience. It wasn't an overnight success; it required meticulous planning, consistent effort, and, sometimes, the strength to say no to immediate pleasures in favor of long-term gains. These skills proved beneficial in other areas of my life, making me more resilient and adaptable.

Going off the grid gave me time to do things that offset the need for certain purchases. Instead of buying all my foods at the grocery store, for example, I was able to grow items in my garden that really added up to a lot of savings and much better-quality food.

Similarly, after making the leap to solar, I was able to avoid having any power bill at all. The average person spends over $3,000 per year on electricity. Having my own water and septic meant another $1,800 in savings.

You'll find that in your first year or two, things tend to cost more than just going to the store, but dollar for dollar, I find that the elevated quality helps it to even out. After your initial upfront investments, you see a lot of savings and benefits on the long tail of these activities.

There is also the element of stress and its impact on your health and overall well-being. Doing some of these things might break even at best when you factor in all of your time and expenses, but the X factor is your health. What is the value of being less stressed about bills and debt over the long term?

How many fewer doctor visits will you need? What pills will you avoid taking? How many other expensive medical needs would you offset by spending quality time in the garden, working with your hands in the sunshine, and not dreading answering your phone for fear of a debt collector on the other end?

I'm sure we could quantify all of that into real numbers, but we intuitively know that putting in a little sweat equity to become debt-free has tremendous value.

STEP 06   ACHIEVE FOOD INDEPENDENCE

—

Eating is an integral part of our daily lives and vital to our survival, but today, more than any other time in history, we've relinquished the responsibility of food production to others. Specialization means that we end up focusing on a narrow set of skills to contribute to society, and in turn, we hope that the necessities of life will be there when we need them.

Many question the wisdom of relying so heavily on grocery stores for food. Even the most basic review of the system highlights its fragility and the serious disruptions that can happen with the machinations of the food industry. The average grocery store only has around a three-day supply on hand, keeping very little in storage, but instead relying on just-in-time inventory to keep its shelves full.

This three-day number always stood out to me in contrast to this quote by Alfred Henry Lewis: "There are only nine meals between mankind and anarchy." While the grocery execs were probably thinking of trucking logistics and rates of food spoilage, rather than looming social chaos, the coincidence is uncanny.

We see how little resilience our local grocery store has when a big storm is forecast by the weatherman. In a matter of hours, shelves become bare and can't be restocked for at least a few days. If those trucks can't get to the store, it can be challenging to keep your fridge full.

Every Thanksgiving, you watch as every item sells out with no more stock to replace it, despite the grocery store owners knowing for months that the rush was coming. It's unnerving to see this when you know that the store has a vested interest in keeping the shelves stocked so they can earn more profits, yet they can't seem to organize enough trucks and food.

Beyond just the reliability of the current system, many of us have concerns about what is in our food. What sprays are used on the produce we eat? What does meat get treated with during its processing that isn't required to be reported on food labels?

There has been a growing sentiment of distrust as the veil gets pulled back through exposés, news reports, scandals, and social media. Good or bad, right or wrong, an increasing number of people want to grow more and more of their own food so that they can have full visibility into the process.

Growing your own food is a great way to mitigate many of the downsides of modern agriculture, along with improving the quality of the food you eat. In many cases, you might not save much money overall, but the value of the same amount of food goes way up. Where you might only be able to afford standard produce, which relies on pesticides, you can grow it yourself and have quality even better than what is certified as "USDA Organic."

You might aspire to grow a large proportion of your food; you might even want to grow *all* your food. I will encourage you to approach your food independence by slowly, over time, whittling away your overall reliance on the stores. You will likely never get to the point where you grow 100 percent of your food, but you'll eventually get pretty close.

Once I started getting into my own food production journey, I focused on the foods that I most enjoyed of those that could be grown in my climate. I also considered the foods that were the most expensive at the store. For me, that meant tomatoes, apples, herbs, and eggs; for you, it most likely will be different.

I had pinpointed these items by looking at my grocery bill, looking at the things I either bought most often or spent the most on. From there, I had a short list, and I then consulted gardening references to find what grew well in my area.

I had to set aside some things for the time being for one reason or another. At the time, I only had a small patch of ground that I could work with, and some items simply weren't going to be practical for me at the time. I enjoy having a good cheeseburger or steak, but it wasn't practical for me to raise a cow. Instead, I joined a local meat CSA run by a farmer who, I knew, raised his herd close to the way I'd want to.

Every morning I enjoy having eggs for breakfast; I realized that chickens were a great option for the space I did have. I started there, with a flock of hens that was a great way of dipping my toe into the livestock pool without getting over my head.

I decided on tomatoes because I loved them in so many forms when fresh from the garden, but they also were one of the easier things to can in jars because of their high acidity. There were also so many great varieties to choose from that grew well in my region.

Having the ability to grow and produce your own food independently means that you are taking control over a fundamental human need. You get to have greater visibility into the process, and you know your food will be there for you even if the shelves at the grocery store go bare.

You'll also come to appreciate the food you do have, because it isn't a matter of swinging by the grocery store on your way home from work. You'll instead understand the hard work that it takes to have any given item make its way to your plate.

STEP 07   LIVE ON YOUR OWN TERMS

—

There is a traditional path that many people (and I'm one of them) follow from the time they are born. In the beginning, we didn't really know better, so we decided to follow the herd, and for a long time, it made a lot of sense to do so. Today, the way the world works is different for so many reasons, and while wisdom is often baked into tradition, the world is constantly changing—which means we need to adapt to it.

You could interpret "the system" in many ways: societal norms, government, corporate influence, etc. Whatever your version of it, that's fine. The point is that there is a subtle pressure to operate within standard structures that might not actually benefit you.

Some might want to rail against the system at the top of their lungs, but actions speak louder than words. I think it's important to point out that bucking the system isn't a matter of being contrarian; it ultimately is just deciding what is best for you. If, after careful consideration, you realize that the life you wish to build veers from the norm, that's something worth pursuing. If you find that the established path is right for you, that is also something worthwhile.

Taking a path less traveled might not be comfortable at times; there are mechanisms that will try to force you back with the herd, but I've found that understanding why you're taking a different approach can carry you through. There will be moments of doubt, and it will be hard, but by the same token, the payoff will be huge.

An example of this was when I decided to pursue my own business venture as a means to escape being tied to a desk. I wanted a job where I could work from anywhere, on my own schedule, and at the time, working remotely wasn't a common thing. I never set out to start my own business, but since I couldn't find a job that let me be remote, I had to go out and make one: I started my website, TheTinyLife.com.

I maintained my regular 9-to-5 for more than three years, while I hustled to build up the business at night. I would come home from the office, make a sandwich, and work from 6:00 pm to midnight during the week. On weekends I was building my own tiny home on wheels from sunup to sundown. It was grueling at times, but knowing why I was doing it carried me through.

When the company I worked for was acquired by a competitor, I realized that we suddenly had two teams for my department and only needed one; layoffs were coming. In a moment of boldness, I marched into my boss's office and offered to be one of the people to be laid off if she gave me a good severance.

I didn't know what her reaction to be, but in that moment, I saw a wave of relief cross her face as she confided in me that she was dreading the cuts that, as I had surmised, were on the horizon. We ironed out the details then and there, and it was a done deal.

Later that week, I was visiting with my parents and shared the news that I was leaving my job to be self-employed. I hadn't told anyone that my master plan was to start a business and live according to my own terms, because part of me was scared I couldn't pull it off.

This news was met with a bit of shock and some concerns, but overall, my parents were supportive. They didn't fully understand how one could make money from publishing content on the Internet, let alone make a living from it, but I had been earning enough to cover my bills for a few months now.

Fast-forward many years: I had bought some land and wanted to build a house on it. However, the bank was having difficulty prequalifying me because they weren't used to working with self-employed people. They wanted to see a pay stub from a traditional W-2 job. They agreed that I had the financials to support the payments easily, but their underwriter wouldn't budge on my not having a W-2.

It was annoying at the time, but while I looked for a builder, I went out and got a traditional job to appease the bank. The difference was that I did all this according to my plan and for my own purposes. Finally, after a long time of searching, I found the right builder for me, and when it came time to sign, the underwriter happily stamped his approval.

This was one of those ways that the system works to keep you on the beaten path. If I hadn't had my own income, I'd need to get the job to live and obtain the loan, then keep the job to pay for that monthly mortgage. Once the ink had dried, I knew I could keep that job as long or as short as I wanted. It also didn't hurt that while this was primarily a charade I was playing for the banks, it was nice to have double incomes.

You will find moments where it gets complicated, where you'll have to adhere to conventional norms to check the requisite boxes, and see it through to the other side. There will be times when others will judge you, and you might find relationships fade away because you've gone your own way, and that will be difficult to cope with at times.

I've found that those who pursue a life that is authentic to them ultimately find a level of contentment, fulfillment, and happiness that is truly rare in today's society. You will find moments of solace in the darker times.

Keep the faith and see it through.

STEP 08   BUILD COMMUNITY

—

You won't be able to do this alone, nor should you try. Finding others to learn from and, in turn, helping others who come behind you is an important part of the journey.

The current system makes everyday elements of life transactional. You don't know the farmer who grows your food; the clothes you wear were sewn by a disenfranchised laborer in another county; and a cow was slaughtered out of sight so you didn't have to face the reality of its life ending so that you can enjoy a hamburger at a cookout.

We've become accustomed to not knowing our neighbors, to just reporting a car accident instead of stopping to help, or turning the many little transactions of daily life into nothing more than the swipe of a credit card. It's the easy way out, not to get involved, but it also comes at a cost.

Getting things locally, like your power from solar panels and your food from the farmers market, or bartering for things you need, brings humanity back to the equation and helps those around you prosper. You'll have to wrestle with the difficult decisions that are typically externalized, and at night you'll need to be able to sleep with the choices you've made.

It will bring an entirely new perspective to your life and test your values all at the same time. The result is that you become connected to the world around you.

As you begin to build your homestead, you'll also start to bond with neighbors and community members who share similar values. At times, you'll work together on common goals. Other times, you'll find a sympathetic ear from someone who has been there during a tough time or a shared problem.

The Internet is an amazing resource for sharing knowledge and is a community in its own right, but it's hard to beat knowing a flesh-and-blood human whose brain you can pick. Each step of your process will prompt you to learn something new, and the old-timer down the road will be a gold mine if you're able to connect.

This really came into play when I started to scale up my garden, to the point I couldn't do everything myself by hand. I had started out with a small, 100-square-foot plot, but it had slowly grown in size over the years. That year, I was on track to break 3,000 square feet of virgin ground.

The ground I was trying to add to my garden had never been tilled before, and my tiller just bounced off the surface. It barely made a scratch on this new plot I was trying to establish, to the point I was about to give up.

Seeing me struggle to make any progress for hours, an old-timer who lived across the street walked over and introduced himself. He had a tractor that he offered to bring over and make quick work of the space. After turning it over with a plow, he disked the whole area twice, and then it was broken up enough for me to use my tiller to finally get the bed to where I needed it.

That experience highlighted how valuable having a community could be. I had constant reminders throughout my journey, and I soon adopted the attitude of paying it forward as much as I could.

You'll find locals to connect with, too. It will be the lady down the road who keeps chickens, the person who had a bumper crop of apples, or the friend whom you swap seeds with at no charge. You'll find that this way of life is abundant, and it reinforces the fact that you reap what you sow.

Start from a place of generosity and be open to meeting others. You'll find that the more you go local, the more you tend to bump into like-minded people. I have friendships that came about because I just saw someone at the local farm supply or at a class on seed saving, or ran into them at the farmers market, or grabbed a cup at the locally owned coffee shop. You'll find your own people that way, too.

PART II

CREATE AN ACTION PLAN

———

STEP 09   SET YOURSELF UP FOR SUCCESS

—

The idea of taking on a life like this can be exciting and daunting, both at the same time. You want to roll up your sleeves and get started, but it's often the case that you don't know *where* to get started. On the flip side, I often see people who want to do it all, and do it all at the same time. Some of us are ready to begin right now; others are on a longer journey that requires additional steps before they can break ground on their dream.

It's at this point that I suggest taking a moment to do just enough planning to make sure you're going about things in a sensible way. When it comes to this planning, there are two major pitfalls: not planning enough, and jumping straight to action without sufficient knowledge; and getting stuck in endless planning that never gets to action. We want just enough to set ourselves up for success, without falling victim to paralysis by analysis.

Going off-grid can mean a lot of different things to many different people. In the next sections, we're going to help refine what your vision is and then what steps you can take to make it a reality. If you have a partner in life or other people who will be joining you in this journey, invite them to work through the planning stage here.

Too often, I meet a reluctant spouse who isn't on board with going off-grid. I've also met my fair share of couples who like the idea, but they get anxious about it because of the many unknowns surrounding the transition. Sometimes, these important people have valid concerns that stop them from getting on board.

Even if you are excited about your future off-grid together, you might have differing visions of what it's really going to be like. Planning with others is a great way to have important conversations to help understand their views, work on solving problems together, and make sure when you say, "We're going to live off the grid," it means the same thing to your partners.

All this will add up to a greater chance of success, a lot less friction, not wasting money, and often a faster path to what you want. Abe Lincoln said, "Give me six hours to chop down a tree, and I will spend the first four sharpening the axe." That's exactly what we're doing here; we are refining our edge to give us the best shot at achieving this dream.

STEP 10   TAKE STOCK

—

Before we get into the nuances of where we're going to mount our solar panels, we want to take a step back and understand where we are coming from and define some large tenets to help guide this whole process. That typically means taking stock of your life and what is most important to you.

The reason I start here is because, too often, I see people who think going off-grid will be the antidote to all their woes. The truth is that going off the grid is not a quick fix for the things you like least about in your life; if anything, it might initially make them worse.

You want to make sure that you're not running away from your problems, hoping to escape them in an idealized vision of off-grid living. If you're deeply in debt to the point that it causes strain on your marriage and your health, that will still be true whether you work in a soulless cubical in a big city or you're sitting on your front porch looking out over your rural homestead.

Life will never be perfect, but we can improve our circumstances dramatically if we set goals and work toward them effectively. The point is, you want to be fulfilled, no matter the context, and if you're going to make the shift to living off the grid, you'll want to make sure that this change supports your overall vision.

Taking the time now to orient yourself is critical because it sets you on a path to get where you want to be. That typically means that you'll need to wrestle with uncomfortable things and face some hard truths about yourself; it can feel a little bit like a self-run therapy session. At times it can be scary to face your demons, but it's important work.

It's also important to make sure that the plan you craft for your new life supports what you need it to. People can get caught up in what they "should" do, and they don't always take a moment to consider what is right for them. What we want to avoid is a scenario where you are a few years into this grand experiment, only to find that it doesn't suit you, or you are stressed out over taking care of everything around your homestead.

The people who are the happiest living off the grid or on a homestead are that way, not because they live in that manner, but because they have done the hard work on *themselves*. From the outside, we might attribute their contentment to their surroundings, but that would be misattribution. The truth is, they're generally just happy people because they've put in the effort to find themselves and choose a way of life that matches that. They could find satisfaction living in the city or in suburbia—it just so happens that their little farmstead is a particularly good fit for them.

To that end, I've tried a lot of different approaches to taking stock of the big picture, and the level-10 life approach seems to resonate with most people. You can do this exercise in about 30 minutes, and I've found it's the easiest way to home in on things quickly. You'll first need to print out or draw your own level-10 life circle: essentially a set of ten concentric circles that radiate outward, then slice the circle up into seven pieces.

Below is an example; feel free to use this or make copies for you to use personally.

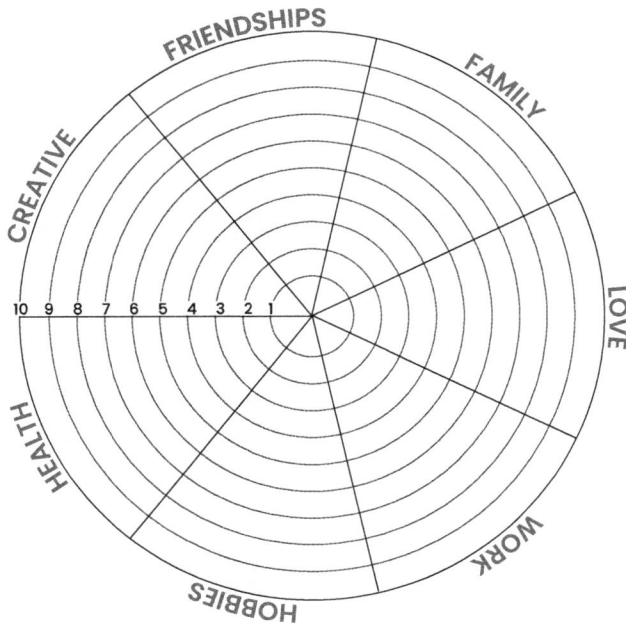

The seven sections map to the key areas of your life; if there is something missing for you, feel free to add it in or to replace one that doesn't resonate with you. Once you've chosen the components, you're going to go through the wheel and rate each category based on where you are today on a scale of 1 to 10.

Your scores are tabulated with 1 representing the lowest satisfaction, and 10 is perfectly satisfied. In our concentric circles, 1 is the circle closest to the middle, and then the scale grows outward. As you shade the wedges of the circles in each step, start by shading from the center and working your way out to the final circle of your rating of that one area.

Ask yourself, "On a scale of 1 to 10, how satisfied are you with this area of your life?"

I find it's best to go with your gut; don't spend a lot of time dwelling on what number it should be but instead follow your instinct. If you give yourself too much time to consider things, you might start to rationalize, which inevitably gives way to kidding ourselves and putting an answer that we wish we had, not what is realistic to the true state of affairs.

Assign your sense of satisfaction a number, and then color in the pie piece up to the level. If you decide that a certain domain is a 4, you'll shade in the area for circles 1 through 4, filling up that much of the wedge. Likewise, if you're a 10, you'll shade in all the sections of the wedge.

Once the wheel is colored in, you'll get a very clear idea of where you are. You will also see how each piece stacks up against the other.

I like this goal-setting tool because it helps you visualize your areas that need the most work. If work is a 3 and everything else is at an 8, then you know that your career needs improvement and focus.

For example, when I did the level-10 life assessment, I realized that I wasn't spending enough time on creative activities. This gave me a clear area I could work on, so I laid out a plan to weave creative activities into my life. I knew it was a weak point for me, and this tool really helped me understand where I stood.

It's unrealistic to expect to be at the tenth level in all areas of your life, and sometimes you need to focus on one area at the expense of another. There will also be areas that are more important to you than the rest.

When looking at my own wheel, I make sure my relationships are always high, and then I try to maintain the others that are in a good spot while dedicating effort to the weakest area in order to bring it up to match the rest. The unshaded sectors of the wheel are places for you to write down what you're going to do in order to improve or maintain certain things.

Maintaining areas that are in a good place (levels 6 to 9) usually only has a few things listed. For example, Hobbies was a level 8 because once a week, I played board games, and most weekends, I went hiking with friends. In that box, I wrote down that I wanted to stay consistent by going to board game nights and going hiking at least twice a month. That got me clear on what I needed to do to keep myself at level 8 without adding anything more because it was already in a good place.

The creative area was too low, and I realized I had been head-down in my work for too long. It was stifling my ability to come up with creative solutions, and it just wasn't as much fun anymore. In my creative box, I wrote that I was going to read one fiction book per month and spend at least two hours per week working on a wood-working project I had abandoned earlier.

You want to get specific with your efforts and make sure that the items you write down for each area directly support your goals in that area. If you're not sure what to write, take your best guess and start there. You'll either figure out that it isn't quite right or that you weren't too far off after all. Regardless, you'll have learned a little bit more so that you can then turn around and decide what to try next.

This exercise only takes about 30 minutes and helps you orient your entire life to the things you find most important. Setting this foundation is the perfect first step in your journey to live off the grid in a way that is right for you.

STEP 11   START WHERE YOU ARE

—

The best place to start is where you are now, and the best time to start is now. If I've learned anything during my time on this earth, it's that there isn't going to be a perfect time; in fact, there is usually no good time to start something. Because of that, you might as well start today.

What I want you to avoid is letting your "someday" turning into "we never got around to it" while you're on your deathbed.

The biggest offender here is waiting until you can buy land to start your off-grid homestead. I've been there myself, wanting 100 acres with a nice pasture, a river or pond on the property, a beautiful barn, and a cozy home. People often have some checklist in their minds, saying, "If I just had ___, I could live off-grid."

Having goals like that is great, but unless you are about to close on such a property, I'd encourage you to start where you are now, wherever that may be. You'll reap immediate benefits that will also help you if those longer-term dreams come true.

Some don't know where to start, so they end up overwhelmed or spending days, months, or even years researching and planning. If you're prone to this pitfall, concentrating on immediate improvements you can make right now is helpful because it gets you out of your loop so you can start taking action.

Finally, the other big stumbling block I see is what I like to call "shiny object syndrome." People get obsessed with the latest and greatest idea, concept, or gadget, spending countless hours learning or tinkering, only to drop it entirely for the next fascinating thing. This tends to manifest in a long trail of half-done projects, without much completed.

All these pitfalls have one thing in common: nothing actually gets done.

The people who are most successful, on the other hand, do just the right amount of planning to know what they need to achieve, and then move to action quickly. You should have a loose road map without every step of the journey planned.

Moving to action is something that can feel uncomfortable or unfamiliar, but if I look at the commonality among the most successful off-gridders—and those in other domains—they all were quick to start. The reason for this is that you don't know what you don't know, and the best way to find out is to do the thing and fail at it.

That brings about another point: become comfortable with failure. You're starting your journey, and there is a lot that you don't know, which inevitably means you're going to make mistakes. That's okay! A quote from Henry Ford is particularly apt when it comes to living off the grid: "Failure is simply the opportunity to begin again, this time more intelligently."

The way I've tried to reframe this process is not as failing, but as asking what I learned. Having this mindset gives you permission to start again, this time a little bit wiser. It becomes a problem to solve, not something you should be ashamed of.

Another commonality of successful homesteaders is they tend to fail a lot but are not bothered by it. Ask any gardener how many times they've had something not work out in their gardens; it happens all the time, but they still end up growing a lot.

The most experienced off-gridders still make mistakes, but rather than beating themselves up over it, they instead keep on going. You're going to need to have grit to make this possible; that means pushing through when times get tough or when you have a setback.

My one word of caution about failure concerns actions that would be catastrophic if they go wrong. Catastrophic meaning the action would likely result in harm to someone or compromising yourself in a way that you likely couldn't recover from.

I guard against these types of risks by asking myself, "Is this action a two-way door or a one-way door?" If it is a two-way door, it means that, if I need to, I can unwind the action and come back to where I was before. A one-way door is a decision that you can't undo without taking a big hit.

This usually boils down to whether you would be willing to lose what is at stake if you were to fail miserably. For me, that usually involves a calculation about money. There is a certain amount that

I'm willing to lose if it doesn't go the way I expected. For each situation, the amount will vary.

A great example of this is buying land. If you are thinking of spending tens or hundreds of thousands of dollars on land, is that a one-way deal, or is that a two-way deal? If the land you found has sat on the market for a year or two, it might be a one-way decision. If, instead, it was on the market for thirty days and you had to make an offer to beat out three other interested buyers, that is likely a two-way decision.

Understanding the importance of moving to action is key to getting started where you are today. Wherever your location and whatever your circumstances, you can begin. It doesn't matter if you are in an apartment in the big city, a rental house in the suburbs, or a tiny plot of land.

For those of you in an urban location, living in an apartment or similar circumstance, it can feel impossible to pursue living off-grid immediately; the temptation is to wait until you can get some land of your own or move into a bigger house with a yard.

Unless you have immediate plans finalized, it's worth starting where you are despite these constraints. You're likely not going to be able to make physical alterations to your housing unless you own it, but there are definitely renter-friendly options that can help you make progress.

An important tenet of all this is that the lessons and skills you learn today are important if you ever want to make the move to a more rural place or further disconnect from the grid. None of your efforts are wasted; in fact, they will give you an edge because you're able to help bend the learning curve ahead of time.

The last point I'll make about starting where you are is that you might come to find that you really *like* your current environment, which might feel at odds with the stereotypical off-grid homestead: an idyllic tract of land in the middle of the woods that has all the rural accouterments that we think are required to go off-grid—a well, a septic system, a solar array, and maybe even a tractor.

Some of the most successful off-gridders are actually those who find themselves in a location that bucks the conventional vision of going off-grid. You might want to be independent and self-reliant, to supply your own power through solar or grow a significant portion of your food, but also really like all that a big city or suburb has to offer.

You might realize that a small lot in a neighborhood is exactly where you need to be to raise your family, surrounded by other families and close to your loved ones. Likewise, you might find that sitting on your porch looking out over your 100 acres represents the quiet you need to live your best life.

Be careful not to get too dogmatic about what it means to go off the grid; be clear-eyed about the strengths and shortcomings of this kind of life. You might find that you can do everything you want to where you are today. You might also realize that the shortcomings of your current location are enough of a hindrance that you want to make a move, but in that move, you'll know exactly what to look for.

When I began my journey, I thought I needed 20 acres to do everything I wanted to do, but after growing gardens and raising chickens for years, I realized that I didn't want to maintain a large acreage—instead, I wanted three really good acres.

Starting where you are will allow you to enjoy immediate benefits, build skills for tomorrow, and develop a lot of practical experience for whatever is next. Let's talk about getting started in different contexts.

STEP 12 UNPLUG IN THE CITY

—

For those who find themselves living in a city, I'd suggest a few practical steps you can take to become more self-sufficient, even in an urban environment with little to no yard. The first is to join a local community garden. There is sometimes a long waiting list, but you can get your name on it, and many garden clubs have social events that you can become part of right away.

Community gardens are a great way to gain access to land you might not have, to help you find a community of like-minded people, and to learn a lot in the process. Community gardens tend to be very affordable; sometimes, they are even free, or you can exchange volunteer hours for any associated fees.

This is how I began my journey. In fact, I ended up starting a community garden myself, since there wasn't one in my local area. I found out that a local church had some land and wanted to start a garden there but didn't know how. I raised my hand to help spearhead the effort, which meant that I now had access to the land. This allowed me and others to start to grow vegetables; we planted fruit trees; and eventually I even convinced the church to let us have a small flock of chickens and quail.

What was even better was that all the tools were on hand to do the necessary work, without us having to buy or store them ourselves. This was really helpful since I didn't have a place for them where I was living, and my car was too small to fit most of them anyway.

We collected dues on a sliding scale, and soon we had enough money pooled to buy a good garden bed tiller for the group to use. I worked with volunteers to trench a water line right to the gardens so we could water our plots easily, and the local nursery donated seedlings and seeds to those who wanted them.

The community garden also opened a lot of connections for its members. Soon, someone with a large kitchen volunteered their house for us to learn how to can vegetables, pickle things, and more. We'd take trips to the local farmers market and do our shopping there together.

For a few days one year when I was first learning, I even carpooled out to an organic farm where we would help the farmer out by working his beds while he answered questions about growing things organically. Most of the time, we'd also leave with a few jars of preserves, a basket of vegetables, or a package of meat as a way for him to say thank you.

Beyond the community garden, you also likely have a local farmers market. The next best thing to growing your own food is to support those who do it locally. Many of these individuals also offer CSAs—community-supported agriculture. These are membership programs where you sign up ahead of time, and the grower delivers a portion of their crops to you each week during the growing season.

These CSAs typically have pickup points, but there are some that also deliver right to your door if you live in a certain area. I had asked the farmer if he would deliver to me directly because the pickup points were farther than I could get to regularly; he said that he could if there were enough customers picking up at this one location.

That prompted me to walk around the office with the sign-up forms, and ten people joined. That was enough for it to be worthwhile for the farmer to make the trip and deliver the produce right to our office. He would drop off his vegetables and organic grass-fed beef.

Beyond just growing your own food, there are other options for going off-grid in an urban center.

A friend of mine who lived in an apartment took his off-grid living to another level when he purchased a solar generator unit that came with three portable solar panels that folded up into a neat little package. On sunny days, he would set the panels up on his small balcony to charge the unit.

This approach could not power all the appliances in his apartment, but it covered a lot of his basic needs. He'd just charge the battery with the panels and use the base to plug in his phone, laptop, and other items.

The unit ended up shaving about 30 percent off his power bill, and he was able to recoup the cost of the unit in about a year.

In addition to saving him money, it also helped him build

resilience, a key tenet of going off the grid. Because he was willing to start where he was and do what he could, the purchase of the panels had the added benefit of giving him a power source when a big storm hit his city and power was down for a few days.

The process also helped him understand how solar worked and gave him practical insights into how much power he used and how long a battery could last under his normal usage. He later moved to a small parcel of land on the outskirts of the city, where he set up a larger system to power everything. His earlier experience made it easier to determine the right size for his new system on the new land.

While he wasn't able to put in a well because he was required to be on the city water system, he did take steps to capture rainwater for his garden and later added a small gray-water system to take the drain water from his sinks and washing machine and pipe it into his landscaping.

Through the entire process, he learned what he actually needed and built a life around his interests and values. Originally, he thought he wanted and needed 20 acres to realize his dream, but after starting where he was, he realized that he could do a lot with a much smaller amount of land.

He right-sized his homestead for himself instead of getting caught up in what he *should* do, saving himself thousands while also keeping him close to the city he liked.

STEP 13   UNPLUG IN THE SUBURBS

—

What I like about starting in the suburbs is that you have some land to work with, but it's often not a lot of space, so it forces you to make use of every inch. The average suburban lot is around 7,000 square feet, just over a tenth of an acre, and this constraint breeds resourcefulness and adaptability.

After my time in a city environment, I rented a house in the burbs, and it was here that I came across a permaculture technique referred to as "stacking functions." The idea is simple in concept, though it takes time to master, but it's well suited to those who find themselves on smaller lots of land.

Stacking functions means designing your space in a way that layers various purposes—in other words, a way that achieves the most bang for your buck. This is a matter of choosing things with multiple functions and combining them creatively so you can do more in a smaller area.

A great example of this is having a chicken coop where you build an arbor over the chicken run. You first have the chickens themselves, a source of eggs and meat. You then plant fruiting vines in the arbor and train the vines to create shade for the chickens during the hot summer months.

The vines grow fruit, which drops into the chicken coop, feeding the chickens and thus reducing your need for feed. Having a good source of nutritious food and being less stressed now that they are shaded, the chickens live better lives and lay more eggs for you. The chickens eat the fruit, and you can use the waste that builds up in the run, after it has mellowed, to fertilize the garden.

You're going to need to plan your yard carefully in order to take advantage of every inch and maximize what you can do within the space. You'll need to think in four dimensions: length, width, height, and time. Let me explain.

When most people plan their gardens, they think of beds that are, for example, ten feet long and four feet wide: two dimensions. But if we then add a trellis for some climbing vegetables, we add the third dimension, height. If we use intercropping to pair a fast-growing plant with slower-growing plants in the same space, we can get two crops out of the same square footage, in the same span of time.

Another example is housing; homesteaders obviously need shelter to live in, but this shelter can do so much more if we stack functions. A house might have a 1,000-square-foot footprint, but be two stories tall: three dimensions. On the roof of the same house, we could install solar panels to capture the sun's power during the sunny months.

We then can layer on the dimension of time because while solar energy does great in the sunny months—reducing our power bill—it struggles during the rainy seasons. If we integrate rain harvesting to capture all the rain of the rainy season, reducing our water bill, we can find savings all year, not just when the sun shines the brightest.

To give you another idea, take the wood stove. We often think about wood stoves as a tool to heat our off-grid homes, but we can do much better than just a heat source. The average wood stove uses only about 60 percent of its heat if it's a modern unit. How could we stack functions here?

What if we also used that stove to heat our water at the same time? Using a coil of copper tubing, we could wrap the copper around the stove pipe and connect it to our water heater. The cool water enters at the bottom of the coil and heats up, causing the warmer water to rise. This then pulls cooler water behind it and starts to circulate water without a pump.

We could also cook on top of the wood stove while heating our house and our water. Nearby we could hang some wet clothing to dry, leading us to use our dryer less and save on our power bill—yet another way we can take the same amount of space and get more from it just by stacking functions.

You're going to find endless ways to make what you have do more, save money, and reduce work all at the same time. Moving from the city to a house with a small plot of land also opened a lot of options for me as I moved closer and closer to being fully off the grid. The main advantage of this new house was that it had more space inside and out.

It was here that I started canning more of my own food. When I moved into this new house, one of the big draws, besides having more land, was a much bigger kitchen. While it is possible to can in pretty much any kitchen, I found the limiting factor in my old city apartment was pantry space. In this new house, I had loads of cabinet space that was begging to be filled with canned fruits and vegetables.

I invested in a water-bath canning pot, Ball jars, lids, and a book on canning. As it was my first time canning on my own and I wanted to make sure I did it correctly, so I signed up for a canning class through my local state extension office.

I canned a bunch of my tomatoes that year as tomato sauce. A farmer at the farmers market had a lot of leftover produce at the end of the day, so I offered to buy a bushel of green beans in exchange for a better price; he jumped at it, and I went home and canned 30 pounds of beans. Later in the year, I went apple picking at an orchard with friends, bringing back enough apples to make apple sauce and pie filling.

All my cooking left me with quite a lot of food scraps, so I started worm composting in a few bins I kept on my back porch. The result was I had this incredible soil from the worm castings; I used it in my own garden and then had so much that I started using it to barter with others.

My suburban rental also gave me the opportunity to start putting the pieces in place to ultimately move to my own plot of land. It was here that I began to build my tiny house in the back corner of the lot. Since it was built on a trailer, I could work at my own pace while living in the rental, and when it was done, I could tow it to my next spot and move in.

Building my own home, even though it was only 150 square feet plus a sleeping loft, was one of the hardest things I've ever done. I was a white-collar worker who hadn't built anything in his life, but now I was undertaking the construction of an entire home.

Despite confronting me with a steep learning curve, it's also one of the better things I've done. Having to learn how to use all the different tools was a huge step up in becoming self-sufficient. Going through the process of building a house from the ground up also meant that I became pretty handy when doing home repairs, wood-working projects, and other skills like basic plumbing and electrical.

Many people want to buy land and build their own homes; some are interested in building tiny, while many want to go full size. Those who have the skills to do so can save about 50 percent on their house if they build it themselves. And if you acquire these skills, it will also give you a massive leg up in all aspects of going off the grid.

It is very comforting to know that I can fix most things if I have to. It's also comforting that, in a pinch, I could find work as a handyman to earn a living if I needed some extra income. The entire experience made me more adaptable, resilient, self-sufficient, and confident in my off-grid journey.

While I didn't get to install solar when I lived in the suburbs— that came a little bit later for me—it's very possible to do on your home since you have a roof to mount the panels to. Depending on your circumstances, you might find yourself in a homeowner association (HOA) where you'll have to work within the rules, but solar is becoming more common, and many HOAs now have provisions for solar installations.

Even if you're not able to put solar panels on your home permanently, you can still adopt the approach used by my friend in the city. Portable solar generators are a great option for anyone and come in many different sizes. Even though I had a 4,000-watt solar array, I still found many times when I would reach for my toolbox-sized power pack for odd jobs around my homestead.

In recent years, we've seen a lot of new entrants into the market, and that has led to prices becoming more affordable, features being added, and quality going up. I now have a small pack that provides around 500 watt-hours as well as a larger one of around 3,500 watt-hours that can connect to my home's power panel.

Small systems are great for when the power might go out, for powering tools to build your off-grid home on raw land, or as a way to save money on your power bill. As mentioned before, I find this small step also gives you the chance to see what's practical when it comes to solar, and this will help inform your purchase of larger systems later on.

Along with solar, you can explore alternative water systems while you're living in the suburbs. Some locations allow you to have septic and wells; this typically comes down to the laws of your town, but some will allow a well for landscape watering and, in some cases, to provide for the whole house. I'll get into the details later on, but for now, just know they might be an option to get further off the grid.

When it comes to livestock, more and more townships are making allowances for people to have a few chickens on their land, even on smaller lots in neighborhoods. I had friends who lived in townhomes with small backyards and had chickens "under the radar," so to speak. While roosters are usually very obvious to neighbors because of their crowing, buying chicken pullets (a young female chicken at least 15 weeks old) is typically a safe option.

It's not that the chickens don't make a sound, but they're generally pretty quiet—especially compared to a barking dog. Having a good relationship with your neighbors will usually be enough to allow you to have a few hens without bothering anyone.

Other options I've found to be good in suburban settings are quail, rabbits, and bees. Bees are very quiet and very subtle; the only way for people to know of their presence is if they happen to see the hive itself, but often hives can be placed in an inconspicuous place, screened by a shrub or fence with little fuss.

Bees are a great way to improve your garden fertility and harvest a lot of honey. The average hive, if taken care of, will produce around 60 pounds of honey. Typically, that is more honey than a family can use in a year, which means you'll have plenty extra to sell, barter, or give as gifts later on.

It's also very easy to split a hive if you have a strong hive and order an extra queen bee. Queens can be purchased widely, and once you have one, you'll take two brood frames (for bees about to hatch) and two frames of honey and pollen, placing all of them in a new hive.

You will then block the entrance to the new hive for about a week; the brood will hatch and adopt the new queen as their own. By the end of the year, you'll have double your hives and they'll both have built back up to full strength.

Quail are another great animal for suburban settings. They are very quiet; at their loudest they'll emit a chirp that can't be heard from even 20 feet away. They also need very little space and don't need to range freely.

I was able to get my first few quail chicks from a breeder in the next town over that advertised online. As chicks, they are about 1.5 inches tall and are about the cutest things you'll ever see. Fully grown, they are about six inches tall and only need two square feet of space to live comfortably.

I was hesitant at first to cage them in such a small area, but after listening to the advice of others, I built a cage that was two feet by five feet and put my five quail in them. I was feeling guilty about what I perceived to be cramped quarters at first, but after watching them for a long time, my fears were assuaged as they seemed quite content.

The cage was also constructed of quarter-inch metal hardware cloth. Raising the quail on wire was also something I was worried about, but I followed the advice of those more experienced. I built the cage with a wire mesh floor so the droppings could fall right through into a compost pile I built below their raised cage.

In the beginning, I was worried that they would be stressed by not having a solid surface to stand on, so I placed a flat piece of wood for them to rest on in one corner of the cage, large enough for them all to fit on comfortably. But after a few months, I ended up removing the wood because they didn't seem to prefer it, being perfectly happy to stand, sleep, and lay eggs on the wire.

These insights make quail a great option for those with small yards. You can have a dozen quail in a cage that takes up very little room, and someone walking close by wouldn't know it's there even if the quail decided to chirp. Quail eggs are great; you'll need 3-4 of them to make up one chicken egg, but it's a great option in places where chickens might not be practical.

Rabbits are something that I don't have personal experience with, but I have friends who have raised them for meat. Two does and one buck can produce 180 pounds of meat per year without much space. The meat is lean, so you'll want to incorporate other fats, but it's an efficient way to grow your own meat.

Rabbits are also animals that don't need a lot of space and are very quiet, too; keeping them in a neighborhood is easy to do and can often skirt HOA rules as they are often considered "pets" instead of "livestock." The space they need is around ten square feet per rabbit, depending on the breed.

In short, whether you are raising livestock or sticking to gardening, digging a well or adding a solar array, the suburban yard can provide you with enough space to do things, but not so much space that you get overwhelmed with managing a large acreage. The space constraints here breed creativity and efficiency, and let you understand what a certain plot of land can do for you.

Despite my own love of a homestead deep in the woods, if I'm being honest, I think most people who go off-grid find the edges of suburbia to be the perfect balance of space, freedom, and access. One to three good acres in a town without too many restrictions is a great formula for many to find satisfaction.

After all, the point of living off the grid is not necessarily to isolate yourself from contemporary society or to eschew all standard systems simply because they're status quo. Instead, we want to find the right balance of leveraging modern technologies and social systems and creating an independent homestead. There are advantages to having stores around when your crops die from a late frost, access to jobs that can pay you a big-city salary while you live in a low-cost area, and a hospital close by if you ever need it. The key, as always, is finding the situation that's right for you or moves you closer to your ultimate goals.

STEP 14 UNPLUG IN RURAL AREAS

—

The pull of a homestead nestled in the woods away from it all is hard to resist. Every time I travel to a big city like New York or London, I have a great time, but after a while, I find myself missing nature. Something about the quiet countryside just feels right.

I don't think this is by chance; just a few hundred years ago, most of us lived in rural areas. Today, in some cities, you are hard-pressed to find trees unless you seek out one of a few manicured parks. My personal feeling is that we belong in nature, that deep down we are genetically wired to thrive in it, but you don't need to take my word for it: the mental, emotional, and even physical benefits of experiencing nature are well documented by scientific studies.

At a basic level, living out in the sticks gives you more freedom when it comes to disconnecting from the grid. You have more options in terms of larger solar installations, wells, and septic systems, which are commonplace in rural areas. You also have more space to grow and raise food, and if, like me, you enjoy your privacy, having enough acreage means not having to see or hear the neighbors.

You might have one acre, or you might have 100-plus acres to establish your off-grid homestead. Regardless, count yourself lucky because many people dream of having a place in the countryside. How you approach living off the grid in a rural location is going to depend on a few things. First, are you planning to rent or buy land? If purchasing, will you buy a raw piece of land (with no roads, buildings, or infrastructure) or it has been developed already? Starting from scratch is a great way to build your dream, to realize the vision as you see fit.

On the other hand, a parcel that already has a house, a well, septic, and cleared land has plenty of advantages: you can move in right away; a lot of the hard work has already been done for you; and there's a lot less uncertainty about whether the land will support a septic system, whether there's water to be had from a well, and whether you could you get power to the site if you wanted grid access as a backup.

Your plans will also need to consider what you want to do on a property. We'll talk about how to figure out what plan is right for you, but different off-grid elements come with different requirements. If you want to have livestock, you'll need pastures, barns, fencing, and more. If you want to have solar, you might need to clear trees to allow for access to the sun. Your gardens and crop fields will need to be cleared, plowed, and amended.

In the upcoming chapters, we'll delve into how to purchase land, develop it, build your homestead, install solar, develop water systems, and more, but for now, I want to share my initial experience of going off grid in a rural area so you can get a sense of what types of activities are possible—even if you're renting land.

Assuming that your ultimate goal is to buy land, renting can provide a stopgap measure. It certainly has downsides—mainly that a lot of the work you put into the property will be left behind—but you don't need as much cash up front. There are also rent-to-own options, in which you rent the land for a defined period at a set price until the deed becomes yours. Renting is not without its risks: often a down payment is required, and if you miss payments or need to move, you will likely not be able to recoup the money you've put in.

In my own journey, I had recently finished my tiny home and found a parcel to rent from a friend. Since I could take my house with me once I found my own land, this option was particularly affordable for me. The land was the biggest I'd even been on; at 28 acres, it had a pond that looked like Walden's Pond, a few open glades, and a mature forest.

The first step to developing my rental property was to get access to it. In my case, there was an old driveway that I had to clean up. I hired someone with a Bobcat (a small construction vehicle) to scrape and level the roadbed. He then cut in drainage ditches on each side of the drive and sculpted the roadbed to be slightly convex so that rain would drain to the sides and into the main ditches. From there we brought in several dump trucks of "crush and run" gravel to lay down a 4-inch-deep driveway bed. It was amazing how fast the driveway came together. A few hours of dirt work, 30 minutes of spreading gravel, and the property was totally accessible.

In addition to the driveway, I had cleared a small area for the pad for my tiny home, parking spaces, and a patio area for a future fire-pit and outdoor washhouse. I moved my tiny home onto the property, added some piers to level the house on the site, and removed the trailer tires to make the tiny house look a little more like a normal house.

The next step was all about setting me up for success in later projects I would develop on this land. I started by building my work-shop so that I could have a place to store tools and to work on those projects. Since I was renting the land, I didn't want to build anything permanent, so I decided on a cargo trailer. The cargo trailer was great because it held a tool chest, a work bench, lighting powered by my smaller solar generator, and the materials I needed. The space was small, but at 14 feet long and 7 feet tall, I could fit full sheets of plywood inside easily.

I could also tow the trailer right to the hardware store, load up materials needed for the job directly into the workshop, then drive to the area of my property that I needed to work on. Having a solid work bench, all my tools, and materials on hand right at the location made doing this work a breeze. A key feature was that the trailer had a fold-down ramp, so it opened the back end up entirely, making loading and unloading simple. I also kept a folding table in the trailer to set up outside the trailer as an extra workspace or a cut station, depending on the project.

My first big project was setting up the solar array in the clearing in the woods. I'll get into more of the details about solar later, but one important element to consider is where to mount your panels. Many people assume that solar panels should be mounted on the roof of their home, but it's actually not the ideal location if you can avoid it. It's more practical when you live in a neighborhood, of course, but where you have more space, consider building an array closer to the ground. That's what I did now that I had room to stretch.

Building on the ground allows you to access the panels more eas-ily for cleaning and maintenance. For me, the real value was real-ized when the first snow hit that year. It was so much easier to clear the snow off the panels when I had two feet planted solidly on the ground, not high up on a ladder.

This arrangement also allows for the panels themselves to stay cooler. The grass below my panels was 80 degrees cooler than my roof; solar panels start to become less efficient as they heat up from the sun. In this way, the roof is one of the worst places to mount your panels because roof shingles can heat up to 150°F!

The one downside to this method is that, if you want to mount your panels close to your home, some might not like the aesthetics of the panels themselves. For every foot of wire you need to connect your panels to your house (and the inverter within it), you'll lose some power to the resistance of the wire.

In my new location, I had parked the tiny home in the shade of the woods and ran the wire to the glade about thirty feet away. To mitigate this loss, I connected three panels in series to total ninety volts to start my voltage high before crossing the wire to the house. This meant that I did lose some voltage, but since I started so high, I ended up with enough voltage to be useable at the other end. This trade-off was more than worth it because my house was totally shaded during the summer, leading me to need a lot less power to cool my home. In this case the savings far exceeded the loss.

There are several approaches for mounting the panels on the ground, but they can be expensive. Of the options out there, I had considered mounting the panels on a single pole with a motor to track the sun to maximize my solar gain. After pricing this out I found that it would require a fair bit of construction, cost, and technical know-how to complete. The cost was also substantial when I factored in the concrete that would be needed in order to mount the pole, as well as the motor to rotate the whole array.

In the end, I realized that the tracking array would increase my cost by about ten thousand dollars and net me roughly 20 percent more power. I compared that to a fixed mount where I would add extra panels; for around 1,000 extra dollars, I would surpass the gain of a solar tracker. So that is what I did. I built my own solar array holders out of treated 4×4's for a few hundred dollars and bought a few extra panels. The whole thing was simple to do myself, has lasted for over a decade, and has the advantage of having no moving parts to break.

In addition to solar, I tiptoed into alternative systems for waste, setting up an outdoor bathhouse in addition to my bathroom inside. If I was unsure about anything, it was my decision to start using a composting toilet. Since I was renting the land, and the local municipality didn't have any codes around composting toilets, I was required to bag the waste in a biodegradable trash bag. Once a week I'd double-bag the contents of the unit and put it out for trash collection. This was by far the least glamorous part of my off-grid life, but soon I got used to it, and it wasn't such a big deal to me anymore. Still, having to explain the setup to visiting guests was a unique challenge.

Alongside the toilet room in the bathhouse was an outdoor shower that I built on a whim. I already had water coming into it for hand-washing, but adding a tankless water heater fueled by a 20-pound propane tank was a massive upgrade. Since I've gone off-grid, propane has begun to take on a more important role in my life. While I wasn't thrilled by the thought of using fossil fuels, it became a necessary element of my off-grid life. Propane is a fuel with a high energy density, and it's also widely available in my area for very little money. In the end, I used propane to heat my water and cook my food, and it was a supplementary heat source when it got really cold in the winter.

Avoiding propane and instead relying entirely on solar would have tripled the size of my solar array and thereby tripled the cost. Each year, I would use about four 20-lb. tanks of propane for everything in my home, which ran me about $45 annually. To put it in perspective, using $45 of propane meant that I didn't need about $35,000 in solar panels and other solar equipment to offset that propane. That's why propane will likely be a key tool in your off-grid toolkit.

Gray water was another big change in my life when I got into more acreage. For the uninitiated, black water is any wastewater that contains sewage, while gray water is wastewater that comes from places like sink drains. It's water that isn't dirty enough to be black water, but it isn't clean enough to be drinking water.

Since I used a composting toilet, I really didn't have any wastewater that wasn't gray water. If I could figure out the right detergents and soaps that would end up in that gray water, I could use it to water plants and even gardens. Previously, this had not been possible on my other locations because I didn't have enough land to spread the water out enough to avoid oversaturating the dirt and ending up with mud puddles.

Since I had so much more room now, I created a small network of pipes that spidered out into the forest that would disperse my gray water over a large area. The water would flow from my drains to a settling chamber where any food particles would collect at the bottom of bin before gradually feeding into the gray-water pipes.

Finding cleaning and personal care products that would not pollute my land was an exercise in researching what was okay to use and then trying a bunch of them out until I found ones that I liked. For dishes and general cleaning, I use Dawn dish soap: it's a great soap that contains only benign chemicals. I also switched to using white vinegar and baking soda for other cleaning tasks, finding it to be a great way to disinfect and deodorize. There were times when I wanted to sanitize certain things and would reach for bleach, but I usually used the bleach wipes for those applications.

Finding a shampoo and body wash was actually one of the hardest tasks for me. I had tried Dr. Bronner's soaps and found that they left my hair looking a little greasy after washing it thoroughly. At the time I was still going into an office, so I needed to find an option that left me looking presentable. After some experimentation, I found a product that worked for me, but unfortunately it's no longer being produced, and I've since moved on to a septic system where I use more conventional products. Just know that you'll likely have to try a few different options before figuring out what will work for you. What didn't work for me, might be right for you.

The final project I worked on was an outdoor kitchen area, which was a great place to entertain but also practical for cooking and when I needed more room to do canning. The gravel patio I had put in served as the base for this and I then set up a firepit, grill, and a cook area with a stovetop.

The signature feature of this cook area was a cob oven, which is essentially a mixture of clay, sand, water, and straw. I built a heavy-duty bench and topped it with fire bricks to support the oven. I started by piling wet sand on top of the table to create a form that would become the inner cavity of the oven. I then mixed the cob on a tarp and started building up the wall of the oven on top of the sand I had shaped for the oven cavity. I added layer upon layer until I had created a continuous dome of cob with a hole in the front to serve as the oven's door. I allowed the cob to dry for two weeks, then came back and applied a final coat of plaster on the outside to give the oven a finished appearance.

Next, I began to scoop out the sand I had initially piled in the middle of the dome. Removing the sand left a cavity that would become the interior of the oven and I let it dry for another few weeks. After the inside of the oven had mostly dried, I built a series of wood fires inside the oven to dry it out further. I started small, then built each fire a little bit bigger than the last until the oven's dome had become a strong, dense shell. Finally, after a few months of work, I was able to cook in the oven for the first time; I started a wood fire and stoked it until it was a nice bed of coals, and the internal temperature had reached over 700°F!

This allowed me to move the coals to one side, scrub the other side of the oven floor clean with a wire brush and use this side for cooking. I typically would bake a loaf of bread first, remove it, and then roast a whole chicken. About 30 minutes prior to pulling the chicken out, I'd slide in a sheet pan of root vegetables to complete my meal (a simple example of stacking functions). Dinner was soon served!

A cob oven may not be part of your personal vision for living off the grid, but I mention it here as one example of the inventive projects you can undertake when you have some space and you've ramped up your practical skills. That's the advantage of becoming more self-sufficient: you can begin to customize your living arrangements, creating systems that fit your unique interests and needs. The scale of your projects partially depends on your acreage, but the general principle is true whether you live in the city or the countryside, whether you're renting your land or are ready to purchase.

How to go about finding and buying land is the next step in our journey.

STEP 15   DEFINE YOUR LAND NEEDS

—

If you're anything like me, you've dreamed of buying land for a long time. There is something about owning the dirt underneath your feet that makes you feel like you're home and all is right. It is no wonder that it's a such a common aspiration for many who want to go off the grid. Unfortunately, land is not cheap and they aren't making any more of it, so it's hard not to notice the massive price tag that comes with such a purchase.

In my own journey, after renting land for years I was able to purchase a property that checked all the boxes. In the seven years that I have owned it at the time of writing this book, it had gone up in value by $70,000—a figure that still shocks me today. The escalating costs mean that for many, land ownership requires years or sometimes even decades of saving.

Exactly how much you need to save depends, of course, on a lot of different factors, including location, but the first determination to make is exactly what *kind* of acreage you need. When I first started looking for land, I was convinced that I needed at least 20 acres of land to do everything that I wanted. Everyone seems to have their own "magic number" when it comes to acreage, but we tend to slip in the phrase "at least" before it. Who wouldn't want hundreds or thousands of acres on which to build out your off-grid life?

I told myself that I *needed* 20 acres because I had visited several farms that were similar to my own vision, and they all seemed to hover around that 20-acre mark. This seemed like a sound approach for determining my desired acreage, but the reality of finding a piece of land that size, and which also included my wish-list items, was tough—and very expensive. I didn't have millions to spend on the land, and that brought me down to earth quickly. I was forced to come to terms with my *wants* versus my actual *needs*, getting very specific and clear about what was a must-have and what was a nice-to-have.

I first started by understanding how big an acre actually is: 43,560 square feet, which seems like a lot of space. But put another way, that's 208 feet wide by 208 feet long; now it doesn't seem all that big. For football fans, a football field is 1.3 acres. Whatever your comparison is, find one that helps you visualize the space of an acre in your mind. I suggest going out to an unobstructed, flat piece of ground and marking out 208 feet by 208 feet and seeing how big it is with your own two eyes.

When you do this, you tend to have one of two reactions: That's way smaller than I realized, or This seems like a ton of space. It's rare for people to react that it was close to what they had in their mind's eye. This is a crucial step to help center yourself on how much land you need.

From there, I started with a blank sheet of paper and wrote down all the things I wanted to build on the land that took up the physical space. You'll have your own list, of course, but to give you a sense of what you might include, here was mine:

- Two-story farmhouse
- Two-car garage
- Driveway with two outside parking spots
- 15 raised garden beds, 4×10 feet in size
- Chicken coop and run
- Area for two pigs
- A workshop, 40×60 feet in size
- Root cellar
- Ten-tree orchard
- Two beehives

The other space consideration I had was that, whatever the dimensions of the parcel I ended up with, I did not want to see or hear my neighbors. I also wanted a plot of land where I would not see my workshop from the house.

For many, your wish list will be very different, particularly if you want to get into larger livestock or do large-scale crop farming. Figure out the things that will take up physical space and write them down. Two other space considerations you'll want to make are whether some of these things can be done in the same space, sharing rather than requiring a dedicated area, and how much space you need between elements. For me, I wanted enough room to have a driveway come up to the house, and for aesthetic reasons I didn't want my workshop to be right next to my house.

A great example of sharing spaces is rotational grazing. Instead of having dedicated fields for each animal, could you develop smaller paddocks and rotate animals through them in succession? For instance, you could rotate cows, sheep, and chickens, along with a rest period. You can divide your pasture into four equal-sized areas. In the first area, start with your cows, who prefer taller grasses and tend to eat only the tops of them. Just before the cows finish off all the tall shoots, move them into the next area and replace them with sheep, who prefer to eat low-growing grasses.

Allow the sheep to graze until they've eaten most, but not all, of the lower grasses, then move the cows to their next area and have the sheep move into the area the cows just left. Back in the first area, you can bring in your chickens to follow the sheep.

The chickens will begin scratching at the manure left behind by the first two animals, spreading the manure and eating any maggots, worms, or grubs they find. This helps to break down the cow and sheep patties more quickly and remove any parasites that might be present. Chickens are often unbothered by parasites that can be deadly to cows or sheep, so letting them eat parasites such as ticks further improves the health of your other livestock.

Finally, move each animal forward in the same order, where the cows will be placed in the last quadrant, the sheep in the third quadrant, the chickens in the second quadrant, and the first area is left to recover and regrow. Depending on how much space you have and how many animals you have, you can time this so that the initial area recovers fully with a thick stand of new grass before the cows come back around to it.

You'll figure out what is required to have its own dedicated space and what can be shared with others. Once you've finalized your list, write down how many square feet you need for each item. For example, my list went something like this:

- Two-story farmhouse—2,000 ft.²
- Two-car garage—1,000 ft.²
- Driveway with two outside parking spots—1,000 ft.²
- Ten tree orchard—950 ft.²

- 15 raised garden beds 4×10 feet in size with pathways—1,500 ft.^{2}
- Chicken coop and run—800 ft.^{2}
- Area for two pigs—1,600 ft.^{2}
- A 40- × 60-foot workshop—2,400 ft.^{2}
- Root cellar—100 ft.^{2}
- Two beehives—400 ft.^{2}
- Space between house and shop/animals—10,000 ft.^{2}
- Buffer from neighbors—4,000 ft.^{2}

Total: 25,350 ft.^{2} OR   0.58 acres

When I totaled it all up, I was shocked by how small that number was—dramatically less than my original aspiration of 20 acres. The land would cost hundreds of thousands of dollars less, and more options would be available to me because parcels as large as 20 acres were scarcer in my area.

Now that I had determined my absolute minimum, I decided that it was worth padding the numbers to account for things not laying out perfectly. Looking over my numbers, I rounded the total up to one acre. Additionally, over time, things will change, so make sure that you build in a certain percentage of space that's designated for growth or the unknown. I decided to add two acres to future-proof my needs, and after more consideration, that got me to a target of three acres as my minimum required space. An important distinction here was that this assumed that all the land was flat, useable, and accessible. That led me to the mantra THREE GOOD ACRES.

For this number of acres to be practical, it had to be prime land, but I could afford it since it was such a relatively small number. This realization made me look at land a little differently when walking potential properties.

Prime land was flat, useable, and easily accessible, and was worth spending top dollar for. If the land had one or two out of the three desired characteristics, I'd be willing to pay an average price per acre. Finally, if the land wasn't flat or useable, I'd consider that as space to buffer against neighbors or the like. I'd be willing to spend some money on that acreage, but it had to be affordable.

That allowed me to look at a piece of land for sale, and in my mind, I'd count the number of good, not-bad, and buffer acres. This really helped me evaluate land as I walked it, and I could add these three buckets up to see if the price was right and make sure it had at least three good acres.

This might be more important to those like me who found themselves in the mountains, where views are great but flat ground can be hard to find. Many pieces of land for sale in my price range were 50-plus acres, but they only had enough flat ground to barely fit a house pad, leaving no room for gardens, solar, or anything else. The rest would be steep land that, with some serious investment, could be made useable through terracing but wasn't worth the cost to do so.

The second criterion of my rubric, useable, came into focus when I considered a wonderfully flat property that was later revealed to be in a flood zone, meaning that I couldn't build on it. Useable is essential for your bare minimum number. This can factor in things like flood zones, rights-of-way, easements, and the like that make a certain spot non-buildable.

Finally, the land needed to be accessible. It's not going to be worth much if you can't get to it in the first place. This includes larger trucks like a cement truck to pour your foundation, the well-drilling rig, and your own car as you go about your daily life.

This model helped me break down a property in a way that allowed me to quickly assess how it matched to my vision, whether or not the price was reasonable, and whether I should consider it further.

In addition to determining the bare minimum number, I also needed to calculate the likely top end. To do this, I looked at the range of cost per good acre and found that I could likely afford up to 15 good areas acres or 50 not-so-good acres if it was available. That allowed me to set the lower and upper bounds of my land search: a piece of land that was between 3 and 15 or 50 acres.

It's important to note that 50 low-quality acres wouldn't be worth it unless they contained my three good acres. When I did actual searches online, I filtered by properties up to 50 acres and my maximum budget.

Size is one factor when looking for land; this is where must-haves and nice-to-haves come into play. The most important "must have" is that the price of the land must be within your budget. I included in my search properties selling for a little bit above my budget because you can usually negotiate down somewhat, but it's important to know your final number.

A key consideration for your budget is whether you would need to sell your current home, if you own one, to fund your land purchase. There are many scenarios that could play out here, but suffice to say, think about how much you can afford without making a commitment you can't support.

It can be hard to decide how much to truly extend yourself versus how much you should hold in reserve. When I went to close on my own property, I could have paid for the land almost entirely with cash, but I held back about $20,000 as my rainy-day fund, and I was very glad that I did. A small mortgage cost a bit more in the end, but I felt more secure in the whole process, knowing that I could get by for over a year if I had to with that cash reserve.

Many people will need to take a mortgage on the land they purchase, which is a challenge in its own right. An excellent way to go about this is to talk to lenders ahead of time to get prequalified and understand what they're willing to loan you. Remember that banks often are willing to loan you more than it's wise to take.

Not all banks are willing to do raw (undeveloped) land loans; in fact, there are few that will. However, some banks will issue loans on land *if* you are also building on that land, often requiring that the home be completed within 12 to 18 months.

I've found that the best option for finding a bank that will lend for raw land is often your local Farmers Credit Union. These are state or regional nonprofit banks with the backing of the USDA. They are designed to give loans to those who want to do anything related to agriculture or farming, no matter how small.

The USDA backing allows the bank to de-risk the loans to an extent and offer better rates than any average bank could. People with decent credit can often get loans with around a 15 percent down payment. Interest rates are slightly higher on land, so understand that you might pay one to two points more than traditional mortgage rates.

Different loans will often require specific parcel sizes. In my case, a property had to be over 10 acres to qualify as agricultural land. I also had to attest that I had the intention of doing something agricultural to some extent at some point; I was planning on doing something agricultural regardless. Of course, you'll have to determine what's right for you. Once you've spoken to a loan officer and gotten prequalified, you'll have your budget, and you can start looking for land, which is an exciting step in the process. Keep in mind that finding land can take a long time, but as you shop, you can save extra money along the way to add to your budget if you're able to bring that money to closing.

Unless you happen to have the perfect piece of land already, you'll likely need to spend quite a bit of time searching for it. This is where the rest of your list of must-haves and nice-to-haves will come into play.

It can be easy to mistake a want to as a need, but try to keep perspective on your two lists. A large part of this will be determined by your local real estate market and what is available at what price. Most things can be had with enough money, but most of us don't have unlimited budgets.

Developing a list of must-haves can be challenging. The best way I've figured out to go about this is to first develop a list of what you think are your must-haves and then add to the list your nice-to-haves. If you have a partner, make sure you're involving them in this process.

With your completed list of both must-haves and nice-to-haves, start prioritizing them, with items that are the most important at the top and least important at the bottom. This helps you quickly weigh the items against each other, usually making quick work of deciding which is more important. There might be a few things you'll struggle over, which are ranked above the other, but do your best.

This list is now your guide. What is left to do is to decide where the threshold is between needs and wants. Imagine a line that demarcates the transition between needs and wants. Start at the bottom of the list, move this imaginary line between items, and move it up the list.

As you move up the list, ask yourself if the next item above the line is essential enough that if you didn't have it, you'd be willing to abandon the entire thing and not have any angst about the decision. This is an excellent way to decide whether it really is a need and not a want, that you'd be comfortable walking away from a parcel if it didn't have that particular aspect. Here was my list when I started:

Land Search Attributes:
- Within my budget
- Three good acres or more
- Established orchard

- Outbuildings already on property
- Big enough to not see neighbors
- Far enough from highways to not hear them
- No HOA or similar restrictions
- Able to connect to the grid if desired
- 30 minutes or less to a big grocery store
- No kudzu or other invasive species
- Mail and package delivery possible
- Good-enough soil quality
- Zoned for a house in a rural/agricultural area
- Neighbors that won't be a headache
- Has water and mineral rights
- Internet of at least 10 Mbps available
- Decent cell phone signal
- Accessible year-round
- Able to percolate for septic
- 30 minutes or less to the hardware store
- Viable well
- Already has utilities set up
- Trash service available
- Stream or river
- Not in a flood zone
- Far from railroad tracks
- Far from major power lines or gas lines
- Has a documented right-of-way or own access
- At least five miles from things that impact resale
- Already fenced in
- Far from potential house developments
- Not downstream from an industrial site

From there, I went about prioritizing the list for my own needs; yours will be different. Finally, I started from the bottom and moved that threshold line up until I decided that if I couldn't have the things above that line, I would be willing to walk away from the entire idea.

Land Must-Have's:
- Within my budget
- Not in a flood zone
- Viable well

- Able to perc for septic
- Able to connect to the grid if desired
- Accessible year-round
- Has a documented right-of-way or own access
- Three good acres or more
- No HOA or similar restrictions
- Zoned for a house in rural/agricultural area
- Has water and mineral rights
- 30 minutes or less to a big grocery store
- Neighbors that won't be a headache
- Good-enough soil quality
- Not downstream of an industrial site
- Internet of at least 10 Mbps available
- Far from major power lines or gas lines
- Far enough from highways to not hear them
- Mail and package delivery possible
- At least five miles from things that impact resale

Land Nice-to-Have's:
- Big enough to not see neighbors
- Already has utilities set up
- Stream or river
- Decent cell phone signal
- Outbuildings already on property
- 30 minutes or less to the hardware store
- Trash service available
- Far from railroad tracks
- No kudzu or other invasives
- Established orchard
- Already fenced in
- Far from potential house developments

Your priorities may be different, but I did want to make the case for a few of mine to help you understand why they might top your own list—a few things that can really make or break a piece of land in terms of its viability and its resale.

While you may not intend to sell, you should always preserve the option to do so. Remember when I discussed decisions that are a one-way door versus a two-way door; this is where most people make their mistake. When you're buying land, the investment is simply too high not to be able to get out if you need to.

Additionally, life happens. Circumstances change, so will your needs. One excellent example: I knew a couple who lived way out in the boonies on a rough piece of land. They had lived there off the grid for years, and as they grew older, medical conditions crept in that meant they needed to seek medical assistance almost daily to stay healthy. Medical care was not available anywhere near where they lived, so after a while they had to make the difficult decision to move into town, where they could get the care they needed.

There is also a scenario in which you build your off-grid homestead and, after a couple of years, you find it's no longer serving you or wasn't what you expected. Living off the grid isn't for the faint of heart; it's not easy. If I'm being honest, I have to acknowledge that there is a strong case for living *on* the grid, where you don't have to spend days chopping firewood, getting up at 5:00 a.m. to milk the cows, or any of the other things that are involved.

Finally, you might be able to build up your property to a point that you need more land, and that might mean that you'll have to move. The point is, you just don't know, and you're better off having the option to sell and not using it, than not having it in the first place. In short, keep the door open.

STEP 16   AVOID FLOOD ZONES

—

This step in your off-grid journey may seem obvious, but it's still worth discussing. I saw some really good-looking parcels in my own search, but they were all in flood zones. Beyond the safety risk, there are typically rules that block you from building in a flood zone, and if you're going to need a mortgage, they will not allow you to be in what they sometimes call a "Special Flood Hazard Area."

Every single county has a geographic information system (GIS), which is essentially a mapping software that's usually online and totally free, where you can see flood zones of any given property. If that's not available, your real estate agent, county office, or the FEMA website can help you look that information up. I'd suggest you seek this information out yourself instead of relying on the seller, just in case you get a less-than-honest answer to your questions.

One temptation can come when a parcel of land is partially in a flood zone. In some scenarios, you might have a buildable homesite that is higher up, putting it outside of the flood zone. I'm not naive about the realities of trying to make your budget work, but I'd still recommend caution in this case. Weather has been getting more severe over the last decade, and what was once just outside a flood zone might move into that category if a record-setting storm were to come through.

There are also times where a homesite is located in a 500-year flood zone, but not in the 100-year flood zone. This nomenclature can be very misleading, and you'll notice that FEMA and others are moving away from it to avoid confusion, but it's still common parlance in real estate.

A 100-year flood is not a flood that happens once every 100 years, but instead a flood that has a 1 percent chance of occurring in a given year. So, this year, it is 99 percent likely to not happen, but it still has a 1 percent chance of happening. Likewise, a 500-year flood has 0.2 percent chance of occurring in any given year.

The thing that comes to mind is Houston in the years 2015–17, when the city experienced three 500-year floods in back-to-back years. The National Oceanic and Atmospheric Administration (NOAA), the national weather service of the United States, has been consistently seeing unlikely flood events increasing in regularity in the last decade.

You'll need to make sure that you're evaluating the risks for yourself, but beyond that, it is good to know that many home insurance policies don't include flood insurance, and that many home lenders will not lend money to a buyer of your land if it's located in a flood zone. Think about making a two-way decision. You'll want to retain the option to sell should you need to.

STEP 17   IDENTIFY WATER SOURCES

—

While you want to avoid flood zones, if you're going to move fully off the grid, you do need a source of water. After all, water is life. We all know this, yet I've seen people who attempt to persevere when a solid water source isn't available, and almost all of them have one thing in common: they end up quitting. The ones who don't quit tend to have something else in common: they can't quit even if they wanted to, because they put too much into their land and they can't get their money back if they were to try to sell it.

Beyond the immediate needs of drinking, cooking, and hygiene, a reliable water source is crucial for sustaining any form of gardens or livestock. This aspect cannot be overstated. For those aspiring to a degree of food independence, this challenge can derail those plans before they even begin. Without water, the possibility of self-sufficiency plummets, as both crops and animals depend heavily on regular access to water.

If you're fortunate enough to find a property with a spring, pond, or river, count yourself very fortunate. The ideal water source is clean, has good volume flowing through it, and has some vertical drop. With enough volume and vertical drop, you'll not only be able to provide for your water needs—you'll also have an amazing source of power. In fact, installing hydroelectric generators will give you one of the best sources of off-grid power, period.

Power from hydroelectric is great because water flows 24 hours a day, meaning that unlike solar, you're generating electricity even at night. If the water source is significant enough, it will even flow in winter months and produce a lot of power.

These types of generators are affordable and quite reliable, with units costing much less than solar arrays and capable of producing a lot of power with a small footprint. There are environmental impacts to consider in siphoning off water to turn your generator, and permits are likely needed. However, since you're returning the water right away to the same stream, it's often palatable to the powers that be.

Unfortunately, it is pretty hard to find raw land with a great water source these days. Meanwhile, developed parcels with great water have mostly been snapped up, and if they are on the market, they come at a steep price. Depending on your circumstances, that might be within your budget, but just know that these types of properties are a hot commodity when they do come up for sale.

Another possibility for developing water sources is available if your land is sloped in such a way that you can create a swale at a natural point in the topography that gathers a lot of runoff when it rains. In some cases, all that is required is putting in a swale and properly sealing the bed of the pond.

Building ponds isn't for the casual do-it-yourselfer; it's better left to an expert. Constructing the earthen dams, sealing the bed, and handling overflow involves a lot of little details that you need to get right. If you have the option to develop ponds on your land, consider looping in a professional for the project.

Barring the availability of water in cases like those described above, you'll likely need to drill a well on the property for your water supply. When it comes to evaluating a piece of land, it's best to see a well already on the property, but that isn't often the case when it comes to raw land. The next best thing is to have any offer be contingent upon a well being drilled; the cost can be added to the total sale price. Even if the seller won't agree to that contingency, my main piece of advice is to never buy land without at least a well permit in hand—unless you plan to stay on the grid when it comes to water.

For those who live in a place closer to a town or a city, there might be a municipal water line you can tap into. In some cases, the city might even require you to tap in; some go as far as not allowing wells to be drilled at all. The nice thing about city water lines is they tend to be very reliable and safe, while a well can sometimes be a headache, and it's up to you to fix it if something goes wrong.

In the area where I was searching for land, wells are commonly attainable. It's very rare to find a piece that couldn't get water; it may be a deeper well than you had anticipated, but the water does come. My offer was subject to being able to obtain a well and septic permit for the land, in my name, before closing. This is a very common and reasonable practice, but it can delay the process. I had to extend my due-diligence period twice because the county was backed up on doing their "perc tests": a percolation test that measures how quickly soil absorbs water. These tests simulate the workings of a septic system and show whether the land can accommodate one safely.

If I lived in a more arid place or where water rights were a big factor, I would be insistent on getting a well drilled and tested before closing. Those wells can be very deep, and the water may never come, so you want to be sure before you close.

The last thing I want to mention is that people are often tempted to rely on rainwater capture, hoping it can be their primary water source. This, too, is something I'd caution you against. Rainwater capture is a great secondary source of water, especially for livestock and gardens, but it can't compete with having a real well or a city water line.

There are some exceptions to this, of course, but to realistically implement a strategy of rainwater capture you'll end up spending quite a bit of money to properly build large-enough storage tanks to meet your needs. The average American uses between 80 and 100 gallons of water per day; the average American household uses more than 300 gallons per day.

Living off-grid certainly makes you conscious of how much water you use, but even if you pair it down to the essentials—taking quick showers and using efficient appliances—it's not uncommon to use 50 gallons for cooking, cleaning, bathing, and drinking. But that's just for you; you'll also have to account for your family, your animals, and gardens.

By heeding the lessons learned from others' experiences, you can be better prepared for the realities of this lifestyle. Let the first investment you make on your land be securing a reliable water source. This is not just an investment in your property, but in your health, well-being, and the success of your off-grid dream.

STEP 18   MAKE SURE THE LAND "PERCS"

—

Once you identify your water source, you need to turn your attention to the other end of the pipe: a way to handle waste and the dirty water from your drains. A functional septic system is a critical aspect that cannot be overlooked when you're establishing an off-grid homestead.

Unfortunately, a lack of these facilities is another reason I see people quitting, and yet they're imperative should you ever need to sell the property. Every municipality also requires them in order to issue a COO—a certificate of occupancy—which allows you to live in your home. Plus, they're essential for anyone planning to live off the grid in a sustainable and hygienic manner.

Septics manage household wastewater in areas without access to municipal sewer services, ensuring that water sources remain clean and the environment is protected. You'll also want this for your own safety because it ensures that your waste is handled in a way that doesn't contaminate your water supply or other parts of your property.

Before you can install a septic system, your land must pass what is known as a percolation test, commonly referred to as a "perc" test. As mentioned in the last chapter, this test determines the absorption rate of soil for a septic drain field, ensuring that the ground can adequately filter and treat wastewater. If your land does not pass this test, it can severely limit your ability to build or live on the property sustainably. There are sometimes alternative methods, but they're often unavailable or are prohibitively expensive.

One of the primary concerns of a non-percolating land is the potential for water contamination. A septic system relies on the ground's natural filtering ability to clean wastewater before it returns to the water table. Without proper filtration, pathogens and pollutants can enter local water sources, posing health risks. Moreover, building regulations often mandate a passing perc test for a septic system installation, meaning that you might face legal barriers to developing your homestead without it.

Since this is so important, you want to make sure your land percs before you buy it. In this process, you'll also determine if there is enough room to install your well far enough away from the leach field (which acts as filter for organic matter) to make sure there is no contamination. Often, the county will also assess if there is enough room for a backup field, should the first septic field become unviable.

As with a well, if you're considering purchasing land that doesn't have a septic already on it, make your offer conditional on your being able to secure a septic permit. Again, make sure the permit is in your name, has not expired, and is real. In my search for land, I encountered several sellers who told me that I could put a septic on the land; one even went as far as saying they had a permit. But when I checked with the county, a perc test had never been done on that land.

Getting a septic installed is typically pretty easy once you have the permit in hand. I would start with getting your well drilled first, though, because if you need to try a second spot, that might impact where you can put your leach field. Wells are also the more complicated part of the process.

In my experience, the hardest part of the whole thing is finding a person who can do the installation. You can sometimes build your own septic system, but unless you've done a few of them before, it's likely well worth hiring this part out.

In many cases, municipalities won't let landowners perform the work themselves because doing so requires a license. Once I realized this, it actually made it easier for me to find a contractor because my county had a list of all the septic companies that had active licenses. They maintained this list to share with landowners looking to hire someone.

When you do your perc test, the county will send an expert who knows soils, hydrology, and geology. This person will require you to dig test pits, and they'll examine the soil's composition and decide if it can perc or not. When I got my test pits dug, we found that one area had some rock higher up than in other areas, so we started doing test pits in the opposite direction and found deep soil that drained well.

It would be pretty rare to pass a perc test but then find that you're not able to build a septic. I'm sure it happens, but the perc test procedure lets the county root out most potential issues before they issue the permit. It also only requires digging down 3 to 6 feet, whereas with a well you might be drilling for hundreds of feet to hit a small pocket of water.

Establishing a successful off-grid homestead requires careful planning and consideration of many factors, with a functional septic system being among the most critical. Ensuring that your land can support such a system through a perc test is not just a regulatory hurdle but a necessary step in protecting your health, the environment, and your investment.

STEP 19   RESEARCH LAND RIGHTS

—

This step will be more or less important depending on what part of the country you live in. In certain areas, you may not have the "right" to the water on your land, including in some cases, the rain that falls on it. For example, you may not be legally allowed to use a river or stream that runs through your property. Rainwater collection can also be deemed illegal and can potentially result in fines, much to the frustration of the landowner.

Similarly, mineral rights determine who owns the minerals below your property. These rights often come into play in areas that are mined or drilled for oil or natural gas. If you owned the mineral rights to your land and a deposit were to be found, you could profit greatly from it or choose to block it.

This is also the case when it comes to fracking: a process in which companies pump hydraulic fluid into bedrock to fracture it (hence the name) in order to extract the oil or gas more easily. There is a compelling case against fracking, often centering around the contamination of well water caused by nearby fracking operations. If you own the mineral rights, others are often not allowed to conduct this type of extraction under your land.

Keep in mind that mineral and water rights can be sold independently of the land being sold. Even if the land is being sold with these rights, contingencies may exist around when and how they are exercised. If land you're considering buying has any limitations, get familiar with the exact terms including expiration dates, conveyances, and sometimes royalty payments.

Some homesteaders strike deals with gas companies that allow the company to extract natural gas from their land in exchange for a monthly payment and unlimited personal use of the gas. One of my friends, for instance, was offered a monthly payment ten times his current job's paycheck, and he could heat his house for free for the rest of his life. This type of arrangement could be very appealing, but if you're considering it, it's critical to understand exactly what the company is going to do and how it could affect you and your land.

The key, in all cases, is to know your land's water and mineral rights, and to make sure you are comfortable with them. They can have a big impact on your enjoyment of the land and, in some cases, your health.

STEP 20   CONNECT TO THE GRID

Yes, you read that right. Connecting to the grid seems counterintuitive given the subject of this book, but you may want to consider it seriously, even if you your goal is to live *off* the grid. One reason is that if you ever want to sell your property and house, it will make things much easier.

To purchase your property, most buyers will need to rely on a bank in the form of a mortgage. And most banks will not underwrite a mortgage on a home that isn't connected to the grid, so your pool of buyers will immediately shrink. While there certainly are cash buyers, attracting those with financing widens the market and likely means you can sell for a higher amount. It also means you might be able to sell to someone who is just coming up like you once did, and while you'll still sell for a market rate, you can also help pass on the legacy to a new homesteader.

There are added benefits to being on the grid. For one, it leaves the option to use that power if your own solar goes down. It can be really useful to walk down to your electrical panel and flip your transfer switch so your house is powered in the interim. That's particularly helpful for powering lights, tools, and other things that are needed to fix your solar array.

You can also benefit from a technique to utilize the grid as a battery through power rate arbitrage. Doing this means you might need a less expensive set of batteries as a power bank. I'll explore this later, but it's hard to beat dollar for dollar.

There might also be the day when you want to start using the grid or use it in some cases. As you age, you might decide that living off the grid entirely is too much hassle, or you might have medical equipment that needs to be reliable. You might have a welder that, which you keep to do repairs once in a while; welders draw a lot of power. You might own an electric vehicle, which would require an array so big it would be cost-prohibitive.

Whatever the reason, whether you use the grid or not, you're not going to regret having the option. Options make you flexible, help you future-proof your personal situation, and enable you to handle whatever life brings.

When you are considering land, make sure power is available to it; don't just assume that if there is a power pole within sight, then you can tap into it. On my own property, I had a neighbor who wanted to buy some acreage behind my land, but it didn't have power. After he talked to the power company, they determined that the only possible way to get the power there was through our other neighbor's land because of the topography.

But even though the neighbor was willing to grant an easement, something she had no obligation to do, the cost to run the power ended up being very high. The laws required that the power company do the work to run the power lines, and the total would be twice the value of my neighbor's land. They ended up abandoning the project and selling the property at a loss because they did not realize it couldn't have power.

That story highlights two important things. You need to have the right to get power to your land, and you need to be able to afford it. For my land, the previous owner had paid the power company to bring a buried line right to the property and, in that process, he got all the legal authorizations to do that.

Before buying any property, make sure you have the ability and money to access power; as with a well and septic, I'd include it as a contingency on any offer you make for land that isn't already connected to the grid. Remember that the power company will likely state that you're in their service area, but that doesn't mean they're committing to bringing in power. You'll need to have a power company engineer come out to the site and draw up the plans. Their legal department will check all of the rights-of-way, and they'll hand you an invoice.

Only then can you have strong confidence that you can get power to the land. I'd prefer for the work to be done prior to closing—and that the seller is the one invoiced; you can add the cost to the land sale. Just be very careful that access to power is a sure thing, because I've seen people's attempts to connect to the grid fall apart several times, which led people to have to sell the land at a loss. Whether or not you ever make use of the grid, it paradoxically gives you the flexibility to achieve your goal of disconnecting.

And while we are on the subject of connecting, I want to briefly mention your Internet connection, which is essential for many of us. Beyond email, general learning, news, social media, entertainment, and more, a solid Internet connection is required for remote working, or for running your own business (farm-related or not).

Reliable Internet can be difficult to get in some remote areas, but luckily it is getting easier and easier. Between existing infrastructure, government grants for rural Internet, and the next generation of satellite Internet, we have more options than ever.

The best place to start is to talk with your neighbors and see what they use. Barring a local hard-wired cable connection, satellite plans are possible and much better today than they used to be. Avoid plans with Hughesnet and Viasat, which are older carriers that are notorious for bad service and trapping people in contracts. More modern options, primarily Starlink at the time of writing, are the next best option when compared to even hard-wired DSL Internet.

Even when you aspire to a simpler life, Internet access is too important a connection to give up.

STEP 21   PICK YOUR AREA

Now that you've figured out the right size of your future property, developed your list of must-haves versus nice-to-haves, and considered requirements that every landowner needs, you can start searching for properties. This prework can take a little time, but a few hours upfront will save you a lot of effort in the long run and, more importantly, a lot of money as well.

This pre-planning also helps you prevent mistakes that you can't undo if you change your mind or circumstances require a new plan. The piece of land you find will never be perfect, but we can at least avoid making a catastrophic mistake.

Regardless, finding land to buy is a challenge no matter what precautions you take, so first prepare yourself mentally to have patience and grace through this process. It will be frustrating and there will be times where it might seem hopeless, but you'll find a way through if you can just keep faith.

Many people already have a general region or area in mind for where they'd like to live. That can help narrow things down, because there is a lot of space in this world. If you don't have ties to a specific area, you'll want to come up with some attributes of your future home to help guide your search. Here are a few questions to help you think through this process:

- How far from or close to a city or town do you want to be?
- Are there particular places you visited that you enjoyed?
- What considerations around employment do you need to make?
- Is there a certain climate that you prefer over others?
- Where are your family and friends located?
- What stores and services do you want to have access to?
- How far is too far to be from the stores and services?
- What type of animals do you want to raise and how many, if any?

- What space do those animals require to live and graze on?
- What price per acre does your budget require?
- What areas are realistic, given your budget?
- What areas can provide the elements on your must-have list?

The answers to these questions will probably give you a sense of possible areas. If you don't know the area you're considering very well, be sure to spend some time there. Go ahead and rent a nearby Airbnb. See what it's like to be a local—visit during the week, during the weekend, and maybe consider different seasons, too.

While you're there, you should talk to locals who are already doing what you want to do, along with anyone else who can share pertinent information. This might mean going to meet the sales person at the tractor supply or at a seed supplier, swinging by the town hall to ask about zoning for the animals you hope to have, or dropping by a farmers market to chat with vendors and patrons alike.

There might be groups on social media you can tap into, classes to take, or meet-ups that you can join. If all else fails, go where your people are, which might mean grabbing a drink at the local watering hole or restaurant. If you're a remote worker, you might have the option to stay for a whole month during prime season and then again in the offseason. The key thing is to get a feel for what life might be like in your potential area and to talk to future neighbors. You can start your land search during this process, so visits don't have to delay you, but they are certainly worth doing.

In my own search, I was considering two main areas that were about four hours apart. Both looked great on paper. So, I made a point of spending over a week in each area, and what I found were two very different experiences. This gave me more confidence and saved me from making the wrong decision.

In the first city, it was difficult to find the basic supplies that I'd need to do my daily work on my off-grid homestead. I knew there was a lot of tourism in the town, but I hadn't realized until visiting that many homes in that city were second homes and most of the homeowners didn't actually live there. After chatting with locals, I also discovered that the town seemed to be on a downward trajectory; land prices seemed good, but that's because people were trying to get out and they were cutting prices to make it happen. There was a fair bit of crime cropping up from a burgeoning drug problem that led to a lot of home break-ins. Looking at crime reports for the area supported the anecdotal stories that locals told me.

The second town looked very similar on paper to the first, but when I spent time there, it was a completely different experience. The downtown was small, but it was comfortably busy. Locals and tourists alike were eating at the local restaurant, the town had invested in a good police force to help keep the peace, and one weekend while I was there the town hosted an arts festival. I learned that word was spreading that this little town was a top place to visit because it had so many events and fun things to do. But it also was early days for the town's upward trajectory. This meant that prices were still reasonable, but buying would lead to rising land value.

This tale of two cities would have been hard to understand if I had not spent real time in them. There is a difference between visiting and living, but approaching the towns with this purpose meant that I could get a little glimpse of what a permanent stay would be like. In total, I ended up spending a fraction of what it would have cost me had I made the wrong decision.

STEP 22   START YOUR LAND SEARCH

—

For many, buying land means working with a real estate agent. I have mixed feelings about the real estate industry as a whole, but as a buyer, you're not likely to be on the hook for realtor fees. Typically, the seller pays around 6 percent of the purchase price of the property to the agents who represent the buyer and seller respectively, roughly 3 percent each.

While as the buyer you're not paying that fee, you may be able to get a deal on a "for sale by owner" property, since the seller doesn't have to make up for the 6 percent in their price. If, on the other hand, the seller has an agent and you don't, you may be at a disadvantage because some realtors will not work directly with buyers, sometimes going as far as never telling the seller you're interested. This can come in the form of agents not returning your calls, outright refusing to work with you, or other more subtle ways.

Realtors will be quick to point out that they bring a lot to the table for their cut, but in the end, I didn't see that benefit in all the times I've worked with realtors. I think that if they weren't so entrenched in the market, bolstered by lobbying, many people wouldn't see the value, either.

The value I do find, however, is in the lawyers, despite my general dislike of them. Regardless of how you broker the deal, with a realtor or without one, you'll need to hire a lawyer to do the title search, review the contracts, facilitate the sale, and dig into the details of rights-of-way, easements, and ownership.

What I found through my land search was that I had to take the lead, calling my agent to book showings and handle other details that should have been automatically handled for me. Unfortunately, many agents simply aren't motivated enough to check for new listings every morning like you are.

So, whether you work with an agent or not, the result is the same:

you're going to be the one leading this land search. The good news is that, in recent years, a lot of new websites have become available to help you find the right property. Along with standard real estate sites such as Zillow and Redfin, there are websites devoted specifically to land, including Landwatch and Landsearch.

The first step in any land search is to use real estate sites to make sure you have realistic expectations about what your money can buy. Your ability to become a power user of the main websites will help you uncover properties that are right for you. Since you've built your must-have list, start there, and set your filters accordingly. When I did this, I set my parameters slightly higher than what I could afford since I might be able to negotiate down. I did this on acreage size, too.

You don't want to narrow things down so much that a good option might get filtered out, but you also don't want to waste time reviewing properties that aren't realistic for you. With each website you'll find the right balance and fine-tuning as you use it.

Many websites now offer the ability to save searches or filter settings as presets if you have account. Others will let you set up certain parameters and get alerts as soon as a new property hits the market. This lets you fine-tune your settings and save them for future use, making coming back to scan for new properties easier.

When I started my search, I looked at all the properties listed on the website, and then I set my filters to only show properties that were listed within the last seven days or that newly met my criteria, such as when a price drop put it in my budget range.

This made my search easier because I only had to look at properties that were new and met my criteria, which was often a smaller list to work through. Every day, I would pull up the website or app and review anything that popped up, which usually took only a few minutes per day. This made it practical to check every day and not miss an opportunity.

Speaking of time on market, you will likely find many properties that seem like a good deal but have been sitting on the market for a very long time. Depending on your location and the state of the real estate market, sitting on the market for 30, 60, 90-plus days may not be abnormal; at other times or in another location, that would be a red flag.

Land tends to take a bit longer to sell than homes do, so looking at how long homes are listed can give you an idea, but homes are a different type of purchases with a smaller pool of buyers. When I was going through my search, the market was very hot. There were numerous times I would be driving to see a property that was listed just an hour earlier, only to get a call from my agent letting me know it was already under contract. But today, as I write this, there is an identical parcel across the road for the same price, which has been for sale for over a year.

Just because something has been sitting on the market for a while doesn't mean that it isn't good; a variety of factors can come into play. Other times, you investigate and it quickly becomes clear why no one has bought the property.

When you first start your search, you'll have to go through these properties and figure out if a parcel is a diamond in the rough—or just rough. This will take time, you'll be let down numerous times as you uncover the dirty little secrets, but slowly, bit by bit, you'll sift through them to see if there's a good parcel to buy.

I will say that if a parcel has been sitting on the market for a long time, you should be wary. Verify what the seller is telling you. Don't take their word for it; ask to see the records, talk to town hall, and make sure you talk to the neighbors.

During my search, I was seriously considering a property, and I decided to take a chance and knock on the neighbor's front door. It isn't always the best thing to just walk up to someone's house in the country, but I wanted to make sure I wasn't missing anything. When a friendly old man answered the door, I introduced myself and mentioned I was interested in the land. He gave me a wry smile and clued me into why it hadn't sold: turns out, the land didn't have a right-of-way for utilities. The seller knew this, but he conveniently didn't mention it and he was not legally required to disclose it.

When I finally found the property I did buy, it had been sitting on the market for a while, but not too long. When I spoke with a neighbor, she clued me into one important detail, about a right-of-way that was inconveniently placed right where the homesite was.

Looking at the plat map, I could see the right-of-way she'd mentioned, but it didn't quite seem to match up with what I was seeing on the land. There was an old access road that people had assumed was the location of the right-of-way, but when I used the map scale, the right-of-way seemed like it was actually in a different location.

I hired a surveyor to verify this, and it turned out that the old road that people thought was the right-of-way wasn't it. The right-of-away actually ran hundreds of feet away from the homesite, cutting through the woods where you couldn't see it. This little detail meant that the homesite was viable, and it suddenly became a prime piece of real estate.

The owner and many other buyers had just written the land off because they didn't pay attention to the details. But the error became my opportunity. I had the survey officially filed the day after closing; I was happy to take the discounted price.

I realize that I was lucky. Chances are, you'll search through all the older lots and not find that diamond in the rough. Instead, you'll need to be ready to jump on newly listed properties. This is when limiting your search too strictly can hold you back, so make sure you're considering options that might not be your first choice but could end up being the right property for you.

STEP 23 FIND POTENTIAL PROPERTIES

—

Once you've identified a property that seems interesting, it's time to pick it apart and get a real sense of what it is like. Begin with a bit of digital sleuthing before you drive out there to see the property, especially if your market isn't moving fast.

This process will take some time to master, but like becoming a power user of the search tools, you'll also hone your online investigation skills. Start with the listing itself: What information does it include?

Land listings often are often sparse on details, and it isn't uncommon to have a hard time finding the address to make sure you're looking at the right area. Even if an address is listed, you can't always trust that information because the listing agent might have been required to enter a valid address, although, if the land is empty, it might not yet have been assigned one.

The two most helpful things I've found to get around this problem are a map view with lot lines and a "parcel ID" number in the listing text. Those two things will go a long way in verifying that you're looking at the right place.

Many listing websites now integrate data from the counties of each state, which conveniently include lot lines. These are often an option buried in the menu of the map view of these sites. You can toggle this on and get a general feel for where the lines of the property lie. Take note of the shape and placement the listing site gives you, then scroll to find the closest road and write down the road name.

With this information or with that parcel ID, you can go to the county's geographic information system (GIS), which will be your new best friend in your land search. It contains so much valuable information, and almost every county has one. These systems can sometimes be tricky to find on the town, city, or county website, but once you do, go ahead and bookmark it. You'll be using this a lot.

Geographic information systems provide the most reliable source of data because they are official records of account and have as accurate a paper trail as you're going to get. For those looking at land, you'll find the following information within the system:

- Lot lines
- Official acreage
- Topographic map
- Satellite imagery
- Floodplain map overlays
- What a parcel is zoned as
- Owner's names
- How much taxes are on the land
- If the land has back taxes owed
- Parcel ID

First, verify that the information in the listing matches what you're seeing in the GIS. I typically look at the lot lines on the listing site, then compare its shape to that of the GIS. If a parcel ID is listed, see if that matches up as well. Finally, look at the neighboring parcels to check if those look about the same.

I will then bring up Google Maps to see if I can find the parcel and whether a street view is available for the lot. The street view can be accessed by using your mouse to click and hold the little person icon in the bottom right of the Google maps screen. If there is a street view, the road will glow blue; you can then drag the person icon while still holding down your mouse button and drop him in the area you want to see.

Many rural places still don't have a street view, but it is sometimes available for the closest main road. I find it worthwhile to drop the person icon on that road and see what you can at the closest intersection. Once you've dropped into street view, you'll see arrows overlayed on the roads that indicate which directions you can travel to see more. Simply click in the direction you want to go and you can get a good feel for the area as you virtually "walk" down the roads.

This virtual sightseeing has been helpful for me to get a sense of the neighborhood. Many places I considered had great photos on the listing, but once I started exploring with street view, I saw that there was a rough house next door or just down the street that was piled with junk cars and trash. I could also see if there were large power line towers, municipal dumps, waste-treatment plants, or anything else that might turn me off from making an offer.

The satellite photos in more rural areas can be hit or miss on Google Maps, but you'll also find that your GIS will have a layer you can turn on with satellite imagery. Find the best available resolution and study it for the lot you're considering and for the nearby lots as well. Here you can get a lot of information.

For example, you'll be able to see where the lot is in relation to stores you frequent. I would always search for the closest grocery store, hardware store, gas station, and farm supply store, taking note of the drive times to each. If you have kids, consider the distance to the local schools.

Along with these neighborhood basics, I also measure a variety of distances using Google Maps and GIS, both of which include a measuring function. I measure things like the following:

- How far is the lot line from the homesite?
- Are there any narrow sections? How wide are they?
- How close is the nearest neighbor's house?
- How long is the driveway?
- How far are the power poles from the homesite?
- How big are the main flat spots?

Among other measurements, you need to make sure that the lot is buildable. The topo layer in your GIS can help you find the likeliest homesite. Then, you'll want to make sure it's wide enough for a house and driveway.

You'll also want to ensure that the homesite is at least 15 or 50 feet from any lot line; each municipality has specific setback requirements stipulating that you're not able to build right on the lot line. This comes into play when you have a small homesite and the required setback would exclude most of it from being used. I've also seen cases in which the only possible homesite is at a narrow point in the land, perhaps only thirty feet wide. This means that you can't build there because you can't be set back from both lot lines at the narrow point.

The length of the driveway is also useful to know, because if you have to pave or gravel it, then the longer the driveway, the greater the cost it will be. In some cases, the driveway might need to be graded; a long driveway could cost tens of thousands of dollars just to give you access to the land.

The distance to the closest power pole, assuming you have the right to tap it (but don't assume), is equally important, since it can be pretty expensive to run power lines a long distance. Some power companies will run a certain distance for free, but after that, the cost per foot is quite high.

See how large any flat spots are, which can give you a good idea of whether they'll be big enough for you to mount solar panels, plant a garden, etc. Most geographic information systems also have a layer for topographic lines that can give you a sense of how steep or flat a piece of land is without your ever needing to set foot on it.

In addition to the GIS, every town or city keeps a deed book, typically online. Like GIS, this is a treasure trove of information. You can typically find the deed book search near the GIS portal on the town's website, but sometimes you have to dig for it a bit. Once you open it up, you can start by searching by Parcel ID, because that tends to be the most accurate.

As you search, you'll find different "pages" of the deed book listed. I often open each one that seems related just to see what is there. You can then also search by the landowner's name (listed in the GIS), but realize that multiple people can have the same name and that the current owner might own other properties as well. Conducting these searches will take some getting used to, but you can find all the old titles, deeds, plat maps, and other documents recorded about the land.

One additional step you should take is to look up the current owner and find the document that shows their purchase from the previous owner. I write down that name and start searching for them, too. I repeat this process for the property I'm looking at as far back as I can, saving each document I find. This is important because you'll find easements, descriptions, deed restrictions, and other important elements carried by the property that might not be noted directly in the latest documents of the current owner, but they are inherited all the same.

Plat maps are a really helpful tool to consider land, understand the property lines, and identify easements on the property. If you don't see a plat map, and you decide to move forward with the property later, you'll most likely need to get an official survey done prior to closing in order to confirm the land's boundaries.

Beyond GIS and plat maps, another source of information you'll want to vet is the real estate listing's photos. These can be very deceiving. When I was going through my land search, I can't tell you the number of times that photos were framed to exclude major pieces of information. Several times, realtor photos were cropped just enough not to show the high-tension power lines right next to the property; checking the satellite maps quickly revealed the truth here.

There was also an instance in which the realtor must have stood on the land they were selling and took a photo of the neighbor's land over the fence, even though the land featured in the photo wasn't even part of the property for sale.

Then there was the magnificent view that tilted up just enough to exclude the neighbor's 10-foot-tall junk pile, topped with a jet ski. Talking to the neighbors revealed that the junk pile owner would drunkenly climb on top of the pile and shoot fireworks off at 3:00 a.m. almost weekly. The country can have some colorful characters!

You're going to find that photos often tell only half the story, which can be frustrating when you're trying to find a piece of land to call home. The best thing you can do is to leverage all the amazing information at your fingertips to spot the red flags.

Another area of the listing that deserves special attention is the description itself. You'll start to learn the code words that realtors use or the ways that they try to frame negatives in a better light. "Rolling hills" can sometimes mean that a property is near a cliffside; in a nice "community" or "lot #" often means an HOA with restrictions; "where you can hear the sounds of the creek" equals possible flood zone.

The listing description also might "bury the lede," hiding the important information at the bottom of the description. This is sometimes hidden by a button to expand the text, so make sure to read the full description.

Finally, your last check should be to talk with the realtor who represents the land. Most include their information right on the listing, and a quick call can reveal a lot. Talk with them about the property and, at the end of the call, ask:

- What feedback did you get from others you've shown the property to?
- Why has the property been on the market for so long?
- Is there anything else I should know before I drive all the way out there?

If you get a realtor on the phone, you tend to get a more honest representation than through a website. While they might gloss over certain details in an Internet listing to faceless people, they're often not as willing to do the same thing when they hear your voice. The key is to combine your Internet search skills with your people skills so that you can get the best possible sense of the property before you drive out to see it.

STEP 24   PREPARE TO TOUR PROPERTIES

—

After you've done all your online research, it's time to see the property with your own two eyes. Scheduling a showing is a pretty straightforward part of the process. Simply contact the agent on the listing. Some real estate platforms try to position themselves as an intermediary between you and the agent in order to get a commission, but you can get around that delay by searching the agent's name plus "realtor." Realtors want to be found, so you can usually locate their websites and their direct cell phone numbers easily.

When you speak to the realtor, they'll often ask if you're already working with another agent and if you have been prequalified for a mortgage if you're using one. If you have a buyer's agent, they are entitled to a commission even if you contact the seller's agent directly; if you don't have one, the seller's agent can choose to work with you directly to show this property or others in their area. They are also happy to keep the entire seller's fee to themselves.

Their question about a prequalification letter from the bank is a way to make sure you can afford the land and that you're a serious buyer. If you're paying cash, they might ask for a "proof of funds" letter, which you can often get from your bank. Some realtors require this to show a parcel, but it depends. If the market is moving fast, realtors can be pickier about whom they show it to; on the other hand, if the market is very slow, they might not even ask.

During my search, I would review my online tools for interesting new listings, and if I saw something that I wanted to see, I'd reach out to schedule a showing early in the week or the coming weekend. I'd then start combing that area to check for other properties close by that might interest me. This way, I could kill two birds with one stone—seeing properties that I might have been on the fence about earlier. Since I was making the trip anyway, it was good to make the most of it.

Before you head out to a property, make sure to verify the address with the agent, since, as we've discussed, the listing information can be misleading. I'd also suggest bringing the realtor's contact info and any plat maps attached to the parcel. Finally, map out the route to nearby stores, town centers, schools, highways, and other points of interest. You'll be evaluating the land itself but also getting a sense of the area as a whole.

STEP 25   TOUR PROMISING LAND

—

When you tour a property, you should start paying attention before you even get to the land.

In many cases, you'll roll through the downtown on your way: What do you notice? As you get closer to the land, what are the neighbors like? As you come up to the property, do you see power poles? What are the road and driveway like?

Once I got out of the car, the first thing I would do is stop and listen. Could I hear nearby roads, neighbors, and other noises, or was it peaceful? I'd also smell the air; one of the lots that I looked at was stunning, but it was 20 miles downwind from a paper plant and the stench was overwhelming.

I then would first make my way to the homesite and pull out my phone to do a few things:

- Check to see if I received cell signal.
- Open a cell signal speed tester app and see how fast the data was.
- Use my phone's compass to find the sun's path for solar panel exposure.
- Pace out the homesite to get a feel for whether the flat area was large enough for plans.
- Mark the homesite on my Google maps app to find later.
- Take several photos of the site.

From there I would walk the property more broadly, seeing how the land sloped, noticing the vegetation and the soil. Think back to your list of must-haves and nice-to-haves and compare the land before you to see if it meets those requirements.

I would also walk the access road to the land to get an idea of how much work would be needed to establish a solid driveway. I'd make note of how much space there was for a driveway, and whether there were any sharp turns or steep sections that might be difficult to navigate with a longer truck.

Well-drilling trucks are usually around 40 feet long, so you want to make sure that one can make all the turns up to your property. The other vehicles that will be used to build your homestead tend to be shorter, but it's important to consider them too. Concrete trucks being able to access your land is going to be important to build your foundations; larger pickup trucks with 20-foot trailers are important for moving tractors and livestock.

At this point, if you think you might put in an offer, you may want to consider approaching the neighbors. People in the country are often a bit wary of unannounced visitors just showing up, but sometimes it's the best you can do. I would also bring along a piece of paper, so I could leave a note if they weren't home.

If everything is looking good, try to find the corner stakes or pins for the property to make sure that what you think is the property line actually is. Compare the land to the plat map you brought with you—the one you downloaded from the deed book or the GIS. Can you get a sense of how the map translates into the real world? That's where the lot's corner pins can come into play. They're usually left by surveyors and give you a starting point if you can find them.

To really get the lay of the land, you might consider bringing a measuring wheel, a tool that rolls along the ground to measure distance. They come in a few varieties, but I'd suggest getting one with a telescoping handle, but also a large-diameter wheel. As you walk rural property, the ground will be uneven and you'll encounter fallen trees, rocks, and other obstacles.

In a pinch, you can use a GPS app on your phone as you walk to measure certain distances, but just know that this method will only be partially accurate. On larger parcels where the homesite or other key areas are well clear of lot lines or other obstacles, your phone might suffice.

The measuring wheel was important when I finally decided on my own land because I was able to use it to determine that the right-of-way going through my property was in fact in a different location than others realized. Finding the corner pin and measuring out the distance revealed the disparity and let me get the land at a lower cost than it was actually worth.

On small lots, make sure to measure out your lot line setbacks to get a feel for the area you'll have left for building your homestead; that inner area, set back from the lot lines, might not be enough room to do what you need it to do.

One additional thing I like to consider is the slope of the land, specifically around the homesite and any steep parts of the driveway. With your homesite, try to estimate the drop across both dimensions of your home. Under 5 feet, you'll be able to build a crawl space; up to around 15 feet you can have a walk-out basement; beyond that, your home will need a lot of work to support it, adding a considerable amount of cost.

The other important consideration about slope is water management. In my search, I found several good-looking lots with homesites near an uphill slope, meaning that when it rained all that water would trickle downhill directly toward the foundation.

In some cases, the flat area of the homesite was large enough so that there was a good amount of distance between the bottom of that uphill slope and the house; in others, the homesite was dug into the side of it, leaving only a few feet between the slope and the edge of the future foundation.

The ideal spot for a house is at least 20 feet from where any precipitation naturally channels. When grading a home, the rule of thumb is that you want the dirt around your foundation to angle down for 15 feet at a rate of one inch per foot. That means the dirt level at your foundation will be 15 inches higher than the dirt 15 feet away in any direction.

If you find that there is a hill above your homesite, you will want at least twenty extra feet on that side of the homesite, solely for water management purposes. That's because you'll want fifteen feet for the grading away from your home and an extra five feet to install a serious French drain to capture any runoff and direct it away from your home's foundation.

I was recently reminded of this when I toured a home with a friend who asked me to come along and share my thoughts on the build quality. As soon as we rolled up to the house, I noted that there was a very steep-grade hill directly behind the house. We toured the home and when we looked at the backyard, the entire space behind the house was boggy under foot. There were obvious places where the water rolling down the slope had washed out part of the backyard. The whole thing looked like a mess.

When we went back inside to check out the basement, we noticed staining on all the wood, indicating that the basement had flooded at one point. Keep in mind that this was a brand-new build: it had only been here for about eight months.

When we asked the realtor, we could tell she wasn't too keen that we had picked up on this detail. That was when she broached the topic of the house's disclosure document. That slope above the home had caused the basement to flood during a heavy rainstorm and mold had set in. The builder, to their credit, cleaned it up and treated the framing with an antimicrobial spray, but this house was destined to flood over and over again because of that hill.

That's not a good beginning to your off-grid journey.

STEP 26   TAKE LAND NOTES

—

Now that you've explored and considered the land, you need to decide whether you're interested in pursuing an offer. If you think the answer could be yes, before you leave, it's time to make a lot of notes and take a lot of photos.

Even if you're not interested in the land, I'd still jot down a few notes that help you remember why you passed on it. I kept these to see if there were any patterns in why I rejected plots, so that I could ask the agent about them before I went out on other future land tours.

A few things I like to make note of include the following:
• How long is the driveway?
• How wide is the driveway?
• What work needs to be done with the driveway?
• Where are the closest power poles? What's estimated distance?
• What are the dimensions of the homesite pad?
• What is the distance from the closest lot line to the homesite?
• Does the land slope across the homesite?
• What is the soil like?
• What is the location of possible septic leach fields?
• How far is the closest neighbor and can you see or hear them?

Next, I would take photos of the property and then do a slow walking video of the entire driveway, the homesite, and any areas where I planned to put gardens, solar, barns, or other outbuildings. This lets you experience the land later on, and a video, if you don't move too fast, lets you lay your eyes on things you didn't realize you needed to photograph.

The last thing I would do as I was leaving was to explore nearby roads around the property. If I was interested, I'd drive all around to see the neighborhood and check a second time for nearby industrial sites, prisons, junk yards, and anything else that might bring down property value or breed crime.

Finally, I would stop in the downtown of the closest town or city, driving by the stores I might be interested in frequenting, looking at the schools, and getting a better sense of the town itself. These are also places to take notes and photographs. That way you'll have records of your potential surroundings, along with your future homestead.

STEP 27   MAKE THE OFFER

—

If you think you have found the perfect parcel—or at least one that meets your needs—my first piece of advice is to center yourself and not let emotion get the best of you. This decision will be one of the most important points in your journey and it could make or break your vision. Next to building a homestead, purchasing land is *the* key to going fully off the grid.

Even before putting in an offer, discuss with your partner or decide for yourself what elements of the negotiation are deal break-ers—what would make you willing to walk away? If you have a part-ner, coming to a consensus can be tricky. I find it is helpful to sit down beforehand and walk through the budget together. If you can't stick to a financial plan as a couple, you should deal with those areas of disagreement even before touring land.

My suggestion for deciding how much you can afford is to base your purchase price on only one of your incomes if you both work. If just one of you works (or you're purchasing alone), then I'd get even more conservative and stay below 40 percent of your income for the payments on the land *and* the house. This may mean that you get priced out of this particular dream once you consider the realities of your market.

Many people's reaction to this advice would be to eschew it; those are the same people who will also end up in foreclosure. If you can't buy the land and the house for a mortgage payment that is less than 40 percent of your income, then you have an income problem.

It seems facile or even heartless to say, "Just make more money," but at the heart of it, if land and home costs in your market eclipse 40 percent of your pay, you'll have to do exactly that. The usual way to accomplish that goal is to take on a new job that can give you the jump in income that you need. If the income from that job is still inadequate, you'll need a longer-range plan; for example, you might job-hop every two years, maximizing pay increases at each hop. Many books and resources are available to help you through this process, but there is a way. Make a plan, put in the work, get after it.

In my own journey, I was able to buy land after I sold a business and had a lump sum to put down. But while my income from my entrepreneurial activities earned me enough for a loan to build a house, the bank didn't count that money as income because of the tax deductions I took against it.

This meant that I had to bite the bullet, swallow my pride, and get a traditional W-2 job that a bank would look favorably upon. I did just that, returning to corporate America for about five years and doubling my income with three job hops. It was exactly what the bank needed in order to qualify me for an amount that the market supported.

This was tough—the entire time I was working those jobs, I was still running my other companies, which meant that I had to figure out how to manage those teams outside of normal work hours and with less oversight. I also had to swallow the bitter pill of going back to corporate life, the thing I had spent a decade building an income to escape.

My plan was to leave the corporate world again as soon as the ink dried in my land purchase, but I ended up finding a position that worked well for the life I wanted to build, being fully remote, no crazy hours, and low stress. The extra income didn't hurt while I was building out my land, either.

Each month I could put half my extra income into retirement and devote the other half to materials, equipment, and some labor to build out everything I needed for my homestead. This meant that I could make way more progress in a much shorter time.

That's what your future could look like if you solve for any gaps in pay: finding a situation you can easily afford and thrive in. Be patient and make calculated moves.

Once you're clear about what you can afford and where your spending thresholds are, it's time to make the actual offer. If you have a realtor, it's a good idea for them to look at "comps," or comparable properties. Typically, they will compare three properties as close to your location as possible, ones that have similar attributes and have sold recently.

Location is a key factor to this, as well as size in acres, with a distant third being features (how level the lot is, how developed, etc.). The average sale price of these three properties is likely the true value of your land. In a pinch, you could look at larger or smaller lots and calculate a cost per acre, then extrapolate it to the land you are interested in.

That price you've determined should be close to the list price, since the listing agent also went through the same process. At this point, you'll need to make a judgement about your offer depending on how hot the market is and how long the lot has been on the market.

If the market is hot, the lot has only been listed for a short while, and the seller is apt to get multiple offers, you will most likely need to offer close to or possibly above the asking price. Realize that other factors like requests, contingencies, and time to close will also be a factor here.

If the market is slower or the property has been on the market for longer, you will have more opportunity to make a deal with the seller. If you notice that they've dropped the price in the past or put it on the market and took it back off for a while, you should be able to get a deal. In such cases, you should definitely make an offer that is lower than asking, but realize that you still need to be realistic by taking the comps into account.

Real estate agents often play the game of pricing their property slightly higher than its actual value and making concessions to certain price points. They might also not want to cross key thresholds, like increments of $50,000 or $100,000, because of how search functions are set on real estate websites. For example, the seller of a parcel of land that is worth $190,000 might consider listing it for $205,000, leaving some room for negotiating down, but they also might not want to cross the $200,000 threshold, so they set the price at $199,000 to get the buyers who are shopping in the under-$200,000 price range in the search filters.

Along with specifying a specific dollar amount, you'll want your offer to include contingencies regarding well, septic, and power if the land does not have them. When I made my purchase, the land had access to power, but the permits for the well and septic had expired.

Since the previous owners had been able to get permits in the past, it was likely that I could get them in the future, but I still made my offer contingent on being able to secure a new permit for both in my own name. Because that takes some time, I asked for an extended "due diligence" period of 60 days.

The seller and I went back and forth on the price twice, and we ended up right in the middle between their asking and my initial offer. Then the deal was signed. I was now officially in due diligence: a period in which the owner has committed to selling to you, and the deal is pending—assuming that you don't discover anything that qualifies as a circumstance that would prevent you from moving forward with the purchase.

STEP 28   DO YOUR DUE DILIGENCE

—

Due diligence is a period during which you are afforded the opportunity to dig deep into the finer details of the property. Unfortunately, it's also the time when many people shirk their own responsibilities, only to end up in trouble. Make no mistake, you want your realtor to do their part and you want to hire a lawyer to research the deed and title, but you also need to take a proactive approach in this process.

I found that most realtors are torn between closing the sale and bringing to light any troubling information. Meanwhile, lawyers are often too busy to do much digging, which means that they do just enough to cover their own hide. It's important to be realistic here about people's motivations and interests. If a lawyer or home inspector constantly flags issues that hurt or derail sales, the real estate agent is less likely to work with them again in the future.

The truth is that professionals involved in land sales, including your lawyer, home inspector, surveyor, etc., have a financial interest in *not* finding certain details. Of course, they are bound by ethics boards, their own morals, and laws concerning their duties, but they also know how to play the game to keep their hands clean just enough.

I raise this point to ground us in the reality that professionals involved in real estate work together on hundreds of deals, each of them getting paid in the process, and each of them earning a livelihood from the process. Let's not be naive about these facts and instead make sure that we advocate for our own interests in this process, too.

That's where you come in. After all, no one cares more about the outcome than you! If you're able to unearth important information that might sway your decision, you can ask your real estate agent, lawyer, or home inspector directly, and they are obligated to be truthful in their responses.

Just pay careful attention to the wording being used. Sometimes professionals involved in real estate deals will phrase their answers in specific ways that are technically truthful, but they may not come out and say what they really think. I am not trying to attribute malice where there is none; someone could feel that they're answering your question, but you did not ask it in a way that demands the full answer you were seeking.

The best way to overcome this is to ask follow-up questions to clarify things. Think through the scenarios that are likely to come up, and lay them out, asking about each part in turn. If your agent or anyone else involved is choosing their words carefully in order to avoid saying something directly, they usually start to get caught up in the misdirection.

I prefer to have these conversations in person in order to get a read on the individual. That way, you hear their tone of voice, see their body language, and can pick up on other cues. If they seem evasive, you can always hit them with: "I feel like there is something important you're not telling me." It's blunt, but it also usually forces them to show their hand.

Regardless of who is involved in your deal, you need to verify a few things during the due diligence period:

- Is the person selling the land actually the owner?
- Do they have the right to sell the land in the first place?
- Are there back taxes, liens, or other encumbrances?
- What restrictions are included in the deed?
- Are there HOA requirements or other rules?
- What building codes govern the property?
- Is the land zoned for what you want to do with it?
- What is the historical paper trail of the property?
- Do you have rights, easements, etc., to access the property?
- Do you have rights or easements for utilities?
- How much are the taxes on the property?
- Is it possible to get mail, trash pickup, and Internet on the land?
- When was the last survey done, and can you get a copy of it?

In some cases, you might need to bring a surveyor out to the property to verify where your corners are and mark the property lines. This can come up if your property hasn't been surveyed in a while or ever in modern times.

When I dug back through the records of several owners of my own property, the original sale only plotted out the property lines using vague landmarks, rather than surveying it properly. The most accurate description for my lot line was, "Beginning on a large chestnut oak near spring on south side." This description of a specific chestnut tree hadn't been written all that long ago, but it was long enough for the tree to have fallen down and rotted into nothing.

Surveys aren't cheap; often they can cost a few thousand dollars and can quickly grow from there. If the land hasn't been surveyed with modern equipment, with a good map created, you can include this as a contingency of your offer: the seller must commission a survey at their own expense.

Make sure to talk to your bank and your town hall about what they will require to buy the land and later build on the lot. You'll want to take that information with you when you get quotes from surveyors in your area.

People generally assume that a fence line indicates the property line. That can often be the case, but in rural land things can get fuzzy over the years. You might find that the neighbor's fence is actually on your side of the line, or an entrance is not where you imagined it to be, or there's something else unexpected.

If an issue like this does arise, you want it to come up in your due diligence period when you have the option to negotiate or even just walk away. If there are discrepancies, lead with diplomacy in mind. A friendly conversation with a neighbor can begin with "The surveyor just sent this over and I wanted to see what we could figure out together."

I've seen this go a few ways, but not every case is a big deal. I had a lot line issue come up on another property when I found that my patio's corner crossed over the line by a single brick. I just trimmed off that corner of that one brick—problem solved.

If a larger area is in question, my advice is to be fair but also firm. You might go back to the seller and use this point as leverage to get a lower price. You could also offer to sell your neighbor a portion of the land (or to buy it from them) and have the lines resurveyed. You could also offer to swap some land to offset the encroachment.

My friend did this when he discovered that his future neighbor's fence was built way beyond the official line, and it was going to cost a lot of money to move the fence. His neighbor was not using a back corner of his lot, which happened to be a perfect homesite tucked into the woods. Both spaces were roughly the same acreage, so they swapped land and split the cost to resurvey.

After your due diligence, it's finally time to sign all the paperwork. Just make sure that your questions and concerns have been resolved well before this. If they haven't, and yet you have been proactive throughout the process, do not feel pressured into closing. Simply tell your agent and lawyer that you need to resolve this issue prior to closing, and they'll be very motivated to make it happen.

Hopefully, on the day of the signing you'll have heard from the lawyer that the title search has been done and no issues found, that title insurance is secured, and nothing is standing in the way of the sale. Then it's just a matter of signing the documents, handing over any checks needed, shaking a few hands, and you're done.

The lawyer will then send the paperwork over to the county office to be recorded in the official deed of records. That can sometimes take a few days, but once everything has been recorded, the land is officially yours!

The last step I'd recommend is to go through, download, and organize every single document you can find about your land. You definitely want to save your deed with the county stamp, your loan documents, plat map, and any signed contracts. I also saved all the old owner records from the deed book, older maps, and any other related documents, which saved me time later. File a paper copy and also save it digitally with a backup.

Do this now, while you have the attention of your realtor and lawyer. It can be easy to get wrapped up in the excitement of buying your new land, but let's have a strong finish to this long journey.

At last, it's time to celebrate and dream.

STEP 29   ORDER YOUR OPERATIONS

—

Now that you have your deed in hand, you're ready to dig in and start making your off-grid homestead a reality. Where to begin? You might remember something very useful from back in your school days in math class—the order of operations, a sequence that you'd follow to make sure you solved an equation correctly. The same goes for off-gridding, which also has a sequence to set you up for success:

- Access
- Water
- Septic
- Power
- Shelter
- Workshop
- Infrastructure
- Outbuildings

Each of these items is listed in order of importance and you shouldn't jump ahead before squaring away each of them in turn. This order also helps you limit your risk because you start with the things that are most beyond your control before moving on to more expensive things that are more within your control. They are also ordered to make sure you can get financing should you need it, or be able to sell successfully.

With the first four items, I suggest not setting up anything temporary, but instead fully building out the resource. So don't just set up a walking path; put in a well-graded driveway. Don't haul in water; drill a proper well. Don't just have a composting toilet; have a septic dug and inspected. Don't just buy an off-the-shelf portable solar generator for the time being; bring in grid power until you have a comprehensive solar array set up.

Your shelter may be a temporary one. Of course, many people want to build a larger home, so a camper or tiny house that you'll live in for 12 months while you build is good enough, assuming that you have all the other elements in place. I wouldn't do this unless you have the financial resources to actually pull off building a permanent home, but if you do, this will work well for the time being.

Next, having a workshop will make everything else that follows it much easier. That way, you have a proper space to house your tools and materials, and a good place to get stuff done. You'll be able to move much faster, save a lot of time and money, and save yourself a lot of headaches. When I built my first home, I didn't have this space, and it meant that about one to two hours of my working time each day were wasted on setting up outside and breaking down at the end of the night.

Infrastructure is bringing power, water, fencing, and sometimes data cables to the place where you need them. We start with these because it can be hard to trench later when you have to work around structures, animals, etc.

Lastly, outbuildings: this is shelter for animals, equipment, and materials that need to be stored closer to the point of use. Not everything can be housed in your main home, nor would you want it to be.

With this order of operations in mind, you can get started on the basics.

STEP 30   START WITH THE BASICS

—

It can be exciting to start your journey once you've purchased land, but first you need to figure out where to start. There are a few key tasks that I recommend pushing to the top of your list, because they are foundational to getting set up on the land where you then can begin your projects.

The first task of developing any parcel is gaining access to it. Access allows you to bring in materials and equipment, and, of course, it makes it possible to get in and out of your home. Unless you have heavy equipment and know how to operate it well, this is often better left to a professional.

In developing driveways for my first and second properties, I found that renting a Bobcat or similar piece of equipment ended up being more costly than hiring someone who owned one of these machines. Plus, they were more skilled at this work than I was, since it was their trade.

Simply put, I could do it, but it would take me twice as long as a professional, and when you're renting a Bobcat by the day or hour, that extra cost eclipses the cost of hiring it out. You also don't need to worry about transporting the machine, fueling it, dealing with breakdowns, or any other hassles. It's also hard to beat the skill of an operator who spends most of their time in the seat of their machine, day in and day out.

In the end I found it cost me about 40 percent less to not do it myself. The exception here would be if you have access to equipment for free or cheap because you know someone; the best neighbors are friendly people who keep to themselves, but who also have equipment they're willing to lend you!

Once you have a proper driveway, your second step is drilling a well and getting a perc test. Water is life, and sanitation is critical. As I've said before, if there is one single thing that causes people to quit their dream, it is not having access to clean drinking water and a flushing toilet. This is the reason I prefer to see a well already on the property before purchasing, or making it a contingency of the sale.

But if those precautions aren't possible, you don't want to invest a lot of money before knowing that you can get water, which is why I suggest this as a second step, before any others. You need a road to get into the property, but I only recommend starting there because you need to facilitate getting your well drilled where you need it.

For your well to function, you'll need power to run the pump, so this is a great time to have the power company bring in grid power. This will not only make code enforcement comfortable but also ensure that it is possible to get power lines brought to the land in the first place.

Now, many will point out, we're trying to build an off-grid homestead! Nevertheless, I suggest you still take this step for a few reasons. First, grid power will make your land more valuable and easier to sell, should you need to. Second, if you need financing to build your house, the bank will require you to be connected to the grid even if you don't use it. Third, while you're building your land, having power will be really useful until you can get solar up and running. And finally, if your solar ever has an issue, you have lights and power to fix it.

If your well location can be drilled without building out a long driveway, I'd do that first. You want to know right up front that you have a solid water source, or at least you should get the bad news now. Either way, start here.

Assuming your land percs and your well has a strong enough flow rate (you want at least five gallons per minute), the next step is to install your septic, even before building the house. This is because you won't be able to get a mortgage without a well and a septic. Also, you have less say over where your septic is to be located compared to your house, because your county will often dictate its placement.

One thing to keep in mind is that many towns require inspections for permits for septic fields—and also a designated backup leach field should the first ever fail. These two areas are places where you are not allowed to build and must be left as bare ground, such as a grassy yard or a hayfield.

Your leach field for your septic typically needs to be 100 feet or more from your well. For smaller lots, the jigsaw that you're assembling may constrain where you can build the house. When you factor in the location of the well, the distance required between your septic and the well, the distance required between your house and the leach field, and the setbacks from property lines, you suddenly may find there is only one location to build your home.

These factors are the reason I recommend following your order of operations and starting with the basics. With this many steps, you don't want to be caught flat-footed right at the beginning.

STEP 31   MAKE YOUR LAND HABITABLE

—

With these fundamentals in place, I suggest that you work to make your land habitable as soon as possible. Everyone is eager to make the move, but you don't want to show up on your land the day after the closing and wind up in a miserable, prolonged camping situation. Having a solid roof over your head, clean water to drink, a toilet to flush, and power to run some basics will make it possible to live here long-term. To ignore or shortchange these necessities will make the process stressful and prone to failure.

Many people make the concession of living in a tiny house, yurt, RV, or some other temporary living quarters until they have time and money to build their final dream home. It's a very common story I hear from others going off-grid. I myself lived in a tiny house until I could afford a larger home.

The challenge comes when these temporary living arrangements are either too rough or turn out to be not so temporary. Whether it's due to lack of funds or unexpected bumps in the road, living in less than ideal circumstances for long periods can burn you out pretty easily. When this happens, it's often the first step toward people abandoning their dreams.

I've seen it too many times: a family sets up shop in a small space that's okay but feels more like fancy camping than a sustainable living situation. Money is tight because they spent all their savings on the land, so they don't have enough to start building their permanent house, or the finish line is years down the road. After about a year, the permanent home is not ready and the family reaches their breaking point. They sell everything for whatever they can get, even at a loss, and try to move on with their lives, but it's not easy to recover from the sense of disappointment and failure. It's a sad ending to this journey and not a place you want to end up.

My advice is that if you're going to live in a temporary situation while building your forever home, you need to have the financial means and time available to complete the final house in under 12 months *before* you start. That may mean a mortgage, which some people want to avoid, but will likely be the only practical option for many.

If you're building your own home and you're not a builder by trade, it will inevitably take you longer than you expect. And if you do decide to undertake this massive task yourself, realize that you need to have the time to do so. It's just not feasible to work a full-time job simultaneously. I realize that this opinion isn't popular, but having counseled thousands of people who built their own homes, I've come to see it as a hard reality.

The problem isn't your ability to build the home but the length of time it will take. Juggling a day job likely means it will take you 18 months to two years to complete the house—and that's if you have all your ducks in a row, cash on hand, and a good work ethic.

If you're in a temporary living arrangement like an RV or yurt, that's going to be too long; chances are, you'll grow to resent it. Additionally, temporary dwellings like an RV often start to need major maintenance from daily use at that point, so you'll be putting out fires at your temporary home even as you're trying to build your permanent one.

I had planned to do this myself on my small home—working a full-time day job and devoting my nights and weekends to building. I finished in 18 months, but pulling it off that quickly almost killed me. It left no time for anything else. If you have a family, other responsibilities, or a long commute, it just isn't going to be possible to complete the house quickly enough. I bring this up not to be pessimistic, but because I want you to succeed. It's important to be realistic about how long we can burn the candle at both ends, and the high costs of taking on such an adventure.

The alternative is to involve a home builder (which likely means involving a bank). Or, if you have a partner, to divide and conquer, with one person working on the house, while the other holds down a good-paying job to pay bills and fund the build. Even if you build the house yourself, I'd suggest budgeting to hire tradespeople for certain phases like electrical and plumbing, and setting aside money for a helper on most days.

If you're building a traditional-sized home, a second set of hands is going to be very helpful and make some things possible that would otherwise be impossible if doing it alone. It's also a great way to acquire skills you don't yet have, which will save you a lot of the time you would otherwise spend learning those skills on your own.

This temporary phase can be an arduous part of the process. But once you complete it, you can look forward to the fun project of setting up your dream off-grid homestead!

STEP 32   IMAGINE YOUR PERFECT DAY

—

It can seem a bit surreal to be standing on ground you now own. If you are anything like me, you've dreamed of this moment for a very long time. Take a minute and just sit with that feeling; you've come a long way, you've worked hard, and this is an important milestone. When I closed on my land, I set up a camp chair with a cooler and watched the sun set. It felt great.

Now comes the fun part: building the dream. We've talked about the essentials: getting access to your land, ensuring you have power and sanitation, and avoiding the pitfalls of a homestead deferred. This is the time to return to the reasons you decided to go off the grid in the first place. You've done your advance planning, but now you need to make concrete decisions about the specific projects you want to tackle. This can be a bit daunting, since the possibilities are almost endless.

I think a great place to start is by reviewing some of the discovery you did early in this journey. You want to make sure that whatever you build on your new land will support that vision for your life. To make this more concrete, I recommend an exercise called "My Ideal Day." Sit in a quiet place, review the level-10 wheel you made earlier, then ask, "What would be my perfect day?"

Think through every little moment of this imaginary day. Start with the moment your eyes open. What do you want that to be like? Do you stay in bed or do you get up immediately and go off to do something? Maybe you let the light wake you up naturally, and then you make your way to the kitchen to make a cup of coffee.

I want you to design this ideal day with the realities of life in mind, too. We all wouldn't mind an endless vacation, but it also isn't practical in this world. So, think about how you can integrate those elements into your day in a way that works best for you.

As you picture each moment and activity, piece by piece, jot down notes. For me, my ideal looked something like this:

I wake up without an alarm and stay in bed for about 10 minutes to wake up. I find my way to the kitchen and make a cup of coffee, sitting on a comfy chair and looking out at the view from my land. I take my time drinking that first cup. From there, I shower and change into some comfy clothes, then sit down to do my most important task of the day without looking at my phone, my email, or any chats.

After completing that important item, I have lunch that I cook from scratch in my nice kitchen and take a short walk afterwards. I only check my email after I've finished my walk. Having taken steps to keep email to a minimum in the first place, I then clear my inbox to zero.

I do a few additional tasks for the next two hours, determine the next important task for tomorrow, and wrap up work around 2:00 p.m. Then, I spend most of the rest of the day outside working on my homestead or in nature.

You get the idea. You are going to lay out all the details, envisioning the day you want, not necessarily the one you have today. For me, it's about managing my energy, doing the most important things when I'm the sharpest, and keeping healthy boundaries with things like my phone, email, and chat apps. I also have cut out social media, news, and politics from my life.

Here are some questions to consider if you're having a hard time when you're doing this:

- What are five words I'd describe my current life as now?
- What five words would I like to describe my future life?
- When I think of my current life now, what do I want more of?
- What do I want less of in my current life?
- What were you doing the last time you lost track of time?
- What stands in the way of doing what you love most?
- What needs to change to make room for more important things?

All of these things are intentional and sometimes difficult to achieve, but understanding where you want to go is an important step in bringing that ideal to life. Figure out what is right for you, then make sure that you're building your homestead to support that vision.

A last note of caution: people tend to get tied up in what homesteading off the grid "should" be, but that definition is someone else's version of the truth. You need to decide what it means to *you*. Along with imagining your perfect day, clarify your overall goals.

Here are some questions to guide you in your discovery:
• What is your purpose in starting a homestead and going off-grid?
• What more do you want in your life after making this change?
• What do you want less of in your life after making this change?
• How self-sufficient do you want to be?
• In what ways do you want to be self-sufficient?
• What parts of your life do you want to stay *on* the grid for?
• Will you raise animals, and if so what kind and for what purpose (milk, eggs, meat)?
• Do you want to make money from your homestead? If so, what percentage of your income?
• How much work do you really want to do consistently, over the long term?
• Do you want to be able to get away from your land for vacations or holidays?
• Can you be consistent in your effort and focus for everything you hope to do?
• How much time can you dedicate to working on your homestead each week?

If you are developing your off-grid homestead with a partner or family, this is a great place to get them involved and invite them to answer the above questions as well. Make sure that you're listening just as much as you're sharing your vision. You want to be working toward common goals rather than a siloed list of activities.

STEP 33   PRIORITIZE

—

Sometimes when I'm trying to sort through all the things that I want to accomplish, it can be hard to narrow down or choose. I've found it useful to actually list out the things that I do *not* want do on my off-grid homestead. This is particularly helpful to get on the same page with your spouse, family, or anyone else undertaking this grand project with you.

Begin by making an exhaustive list of what your homestead will not be. Once you've done that, your mind will be primed to imagine alternatives. Look at the items in your "not do" list that you felt most strongly about—and write the exact opposite of that thing in your "do" list. This will help you not only avoid building a life you don't love, but reinforce the parts you'll value the most.

An example for me was raising cows. I didn't want to take this on because it's a lot of work, would mean I'd have to wake up early, would make it harder to go on vacations, and requires a lot of space and equipment. To me, the alternative to raising cows was making sure I had space in my off-grid house for a chest freezer large enough to fit the meat from a whole cow and 50 chickens.

Create your own alternatives: listing out the opposite forms of each item on your "not do" list.

From there, expand your list to other things you want to pursue on your homestead. This can include power generation, buildings, crops, gardens, animals, equipment, activities, income sources, and anything else your heart desires.

Have your family members or partners add their own items to create one giant list. Go wild with your ideas at this stage. Shoot for the moon; we'll sort it all out next.

The last step of this process is to combine any similar ideas if you share the same vision as others. Build consensus around overlaps, and if the other person feels the two things are different enough, spell out how they are different. You'll have a list of everything you *could* do. Now the question becomes: Where to start?

The easiest way to prioritize is simply to compare each entry on the list. Ask whether each is more important to you than the thing above it. If yes, move it up. Then ask again, is this more important than the entry above it? Repeat until you get to "no" or you get to the top of the list, then start with another idea. After doing this for a while, you'll wind up with an ordered list of priorities.

The key thing to remember here is that you're not deciding whether or not to tackle each thing; it's just a question of do we do this now, or do we do this a little later? If you're working through this activity with your spouse or family and you're having a tough time with competing priorities between people, use a weighted vote.

Take your top 10 ideas, and assign everyone 100 "points" that they can allocate to each entry on the list. If something is interesting to them, they can allocate more points to it. If it's critically important to them, they can allocate the majority of their points to it and only a few points to the rest. In the end, total the points from each person for each entry and rank them from highest to lowest.

Suddenly, you have a collective list of priorities—and a place to start.

Keep this list and refer back to it as you begin your projects, only starting the next item on the list when the previous one is finished or is in a holding pattern because you have to wait for something (like the next season to plant, or a part is backordered, etc.).

I'd also recommend some back-of-the-envelope budgeting and not starting anything until you have all the funds to complete it. That may mean you save up while you go on to other projects. Don't begin a bunch of projects that you can't afford and leave them half done.

The reason for all this planning, prioritizing, and budgeting is to avoid "shiny object syndrome," which I mentioned earlier (in step 11). You don't want to spin your wheels hopping from one half-finished project to the next, expending a lot of effort but not making much progress.

This isn't about being productive, industrious, or even obsessed with goals; it is about making sure your dream comes alive and doesn't turn into some half-baked mess. I also find that this approach is more efficient and faster than multitasking or project-hopping. And for me personally, it calms my mind; rather than worrying about a tangle of competing ideas, I can concentrate on one thing at a time. My tasks don't weigh me down because I've decided to focus on a narrow set of achievable activities that I can accomplish in a reasonable amount of time.

PART III

MAKE THE DREAM A REALITY

———

STEP 34   DO THE WORK

—

Now we get to the fun part: doing things.

After all the planning and budgeting, it's time to actually get your hands dirty—which is probably what you imagined when you first dreamed of living off the grid. There is a lot of energy and excitement about getting to this stage: it's when I feel most alive.

The great thing about this work, despite it being hard, is that you're building your dream. You are building your life. That can be a stark alternative to being stuck in a job where you feel like a cog in the machine.

Unfortunately, far too many people don't feel a sense of purpose in their work, so it's incredibly invigorating to be able to redirect your energy to projects you care about and activities that will benefit you and your family. Not much else in life will compare. It also tends to make those parts of life we don't love—like working an unfulfilling job—so much better because you're no longer just working to pay the bills, you're working toward something better. That somehow makes the job easier despite the day-to-day staying the same.

In this part of the book, feel free to jump to the parts you need and skip the things that you don't. Start with the item that is your top priority. Learn what you need to learn right at the moment you need it, and then get out there and do it. Then once you've completed that task, move on to the next.

My advice for every project is to start small and not take on too much at once. Instead of trying to grow all your own food right from the beginning, grow 5 percent of it. Instead of going totally off the grid, start with offsetting your power bill for your fridge or lights. Instead of starting with a large flock of chickens, start with three in the first year, then grow from there.

I may sound like I'm stuck on repeat here, but see each project through from start to finish. Again, I can't tell you how many homesteads I've visited that are littered with a series of incomplete projects—and when I go back, sometimes years later, not much has changed. This may mean at times that you're not doing a whole lot other than saving extra money, working a side gig, or finding another way to finance your next project. The difference is that you're working to save for something concrete, not something abstract that might happen someday.

I had mentioned that in my own journey, I kept that corporate job for longer than I expected in order to bring in additional dollars to get some bigger projects done quickly. The extra cash went straight to my savings and allowed me to complete projects that otherwise could have taken a decade to finish.

You might also consider bringing in contractors or a helper to speed things up. While I encourage everyone to become more self-reliant, there is also a case to be made for hiring an expert or second set of hands if it means finishing a project much faster.

The refrain I hear too often is "We didn't have a lot of money, so we just figured it out." Again, there is a lot of merit to this mindset, but you also run the risk of becoming (to use another common phrase) "penny-wise, but pound-foolish."

In the portion of the book on budgeting for your land, I talked about how sometimes you can't save your way to affording land if you don't earn enough in the first place. I see people trying to make two ends meet when there isn't enough slack in the rope to make it possible.

Part of the solution is to develop skills that are so valuable that you can work fewer hours for the same or more pay. If your skill or trade pays a good hourly rate, you might only need to work part-time, allowing you to spend the rest of your time developing your off-grid homestead. This also allows you to scale your hours up and down to afford the things that you need to do.

I see people spend hours driving around to yard sales, finding scrap wood on local online listings, or coming up with something that is a sloppy solution that they're constantly fixing. There often isn't a lot of intention and thought put into it. When you total up the hours they spent and the gas they used up, you realize they might have been able to work those hours and earn enough to buy the thing in half the time. The more your hourly rate, the easier it is to gain leverage. Developing skills that others are willing to pay for means you can maximize your time and accomplish more. You can also choose to do the work yourself, but it's because you want to, not because you must.

That was the calculus I did when it came to building my second home. I sat down and figured what a builder would charge to complete the project above and beyond the materials. (A common way to estimate total building cost is to double the cost of the materials alone.) I then figured out how much time it would take me to do the same (which would be longer because I don't have as much experience). In the end I determined that I could either do all the work myself in lieu of paying a builder or I could work at my current job for about six months. It was tempting to take on the task, but I realized that working with a good builder was a more efficient use of my time.

The home build was going to take about six months if I used a contractor, or 18 months if I did it myself. Income-wise, I earned about the same amount that the contractor would charge me, so our time was equal to each other.

That meant that I could work 18 months, when it only cost me 6 months. In other words, I was going to be 12 months ahead in dollars in my bank account. Plus, since the contractor was more experienced (and I had vetted them thoroughly), they were more likely to do a better job. I was able to move in sooner, and all the headaches of subcontractors, materials, etc., were their stress to deal with, not mine.

There are trade-offs in every decision about whether to hire out the work or do it yourself. Part of the calculation is determining what activities will overwhelm you and which will bring you satisfaction and a sense of accomplishment. Your decisions may be different than mine. What's important is that the work you do take on fulfills you. After all, that is the central purpose of developing your off-grid homestead.

STEP 35   DWELL, TEMPORARILY

—

Many people want to move to their land as quickly as possible and get to work building things out. This often requires temporary housing while you're building your forever home.

There are many options here, but my favorite, assuming that you live in a moderate climate, is a yurt. For those who live in locations with severe weather, including very cold winters or scorching hot summers, you'll have to establish more robust housing just to ensure you'll be able to make it through these seasons.

For the rest of us, yurts are the perfect mix: spacious enough for you to live comfortably, fairly quick to set up, and a price that isn't out of reach for most. They have good ceiling height, and the dome and side windows are great for natural light. When you set up a yurt, the longest part of the process is building the deck on which the yurt rests, which can take about a week. From there, the yurt can be set up in roughly two days. Local building codes tend to be easier, too, because a yurt is considered a temporary structure, essentially a tent.

The key distinction is that many municipalities might not allow you to officially live there full-time. When I explored this for my own land, I had to get the deck inspected, but then I waited until the code enforcement went away before setting up my yurt. If they ever asked, I simply said it was "storage" for building materials.

I prefer yurts to some other options, including a pre-built shed, tiny house, or RV, because they give you a good amount of space for the cost and they're also very easy to break down once your main home is built. I also like them for seasonal rentals if you decide to earn a little extra income after you've moved out. Because their cost is low, you can quickly earn enough to break even; then you're earning pure profit.

The main downside to a yurt is that they're not very secure from those who might want to break in—including rodents and bugs. But if you build your deck two to three feet off the ground and keep the area cleared and mowed, you mitigate most of these issues.

If you're in a cold climate, you'll burn a lot more wood in a yurt than in a traditional house to keep warm, but it is possible with a good wood stove. For summer, you can cool your temporary home by opening the top dome and lifting the bottom a few inches off the sidewall fabric: this creates a good convection. Whatever the season, a yurt will never be as comfortable as a real home, but depending on the climate and your tolerance for hot and cold temperatures, it might be workable for you.

You do have other options. RV campers can be bought used, but they don't have much more insulation than a yurt and are about three times the cost per square foot. It's important to keep in mind that RVs are intended for seasonal use, which, to the industry, means you'll go camping in them for a few weeks a year. As such, the manufacturers build their RVs to that specification, not for full-time use. With full-time use, low- to mid-cost RVs start to develop problems at the five-year mark, with major issues coming in years seven to ten. Keep in mind that if you buy used, the clock has already started on that RV.

Also, remember that homesteaders and farmers tend to be rougher on the RV itself than typical users. While working on many projects, chances are you'll be coming and going often, using the kitchen more than campers would, and even bringing projects inside to get them out of the elements. This all adds to wear and tear that an RV is not designed for.

Sheds are close in cost, but you then must factor in windows, doors, insulation, and finishing. You can build your own shed, of course, or buy a premade shed to get started sooner. The place I see many people go wrong is not investing in a good pad on which to place the shed and not insulating the floors. Scrape the top layer of soil off, level the site, and then put in a thick layer of gravel and insulate the floor with closed-cell foam.

A tiny house is an excellent option for longer-term living, but the cost is higher than you'll likely want to spend on temporary shelter. If you plan to live in a tiny house long-term, this option makes a lot of sense; just make sure you budget for a proper cement parking pad and some outdoor living space. The real downside is financial; the cost of a tiny house is usually high enough that you are better off using that money as a down payment on your forever home.

You'll need to decide what is right for you, but as mentioned earlier, I suggest that you don't stay in your temporary arrangement for more than 12 months. That's when I find people start to get uncomfortable and dread the next winter or summer. You should make sure that you have a plan and financial means to build your main home within the year to avoid burnout from living in such austere circumstances.

STEP 36   BUILD YOUR OFF-GRID HOME

—

An off-grid home is not all that different from a traditional one; the main distinction lies in the way you power it, provide water, and handle sewage. However, some design considerations can give you an edge when it comes to reducing your power requirements for things like a solar array. There are also usually some additional considerations for pantry, cooking, and other space requirements for homesteading work.

A few key design and material choices can make a big difference in your energy needs. Understanding these factors could save you tens of thousands of dollars up front and over the lifetime of your home.

The best way to tackle a home's energy usage is to understand what parts of that home use the most energy in the first place. If you can make decisions that limit usage in those areas, you can make the job of powering your home a lot easier.

The typical breakdown of power usage in an average home is:
- 50 percent on heating and cooling,
- 15 percent on water heaters,
- 13 percent by your washer and dryer,
- 12 percent for lighting.

Those four items account for about 90 percent of your power needs. Therefore, if we can design a home to minimize these four areas, we do ourselves a lot of favors. So how do you do that?

Heating and cooling, also known as HVAC (short for heating, ventilation, and air-conditioning), is the most significant energy sink for any home, far and away beyond the others. With rising temperatures and increasing severity of storms, a solid HVAC system is important to keep you comfortable, and in extreme weather, the considerations aren't just about comfort, but safety.

You'll quickly learn that it's easier to heat an off-grid home than it is to cool one. Woodstoves, propane, or natural gas heaters all are great options and are easily installed with materials you can find locally. Cooling requires more complex equipment like an AC unit, heat pump, or the like, and they are difficult to power with even large solar arrays.

One major thing you can do to mitigate this expense is to choose a location with a mild climate—not too cold, not too hot, and not too humid. I see many people buy cheap land in places like the Arizona desert. Beyond the fact that you most likely can't get water there, the climate is very harsh, making it difficult to cool your house even if you were on the grid.

If you had to choose the lesser of two evils, I'd opt for a climate that gets colder over a hotter one, because you can easily add a wood-stove for a few thousand dollars, whereas in a hot climate you'll have to triple the cost of your solar system just to power your AC unit.

Keep in mind that with cooling, the challenge really isn't during the day because the sun is out, and your panels can dump all that energy directly into the cooling; rather, it's nighttime. It may be cooler at night in many hot places, but that doesn't mean you won't be running your AC. The difficulty comes when you need to run your AC on a sweltering night, and it's been cloudy for a few days. All that energy comes from batteries that you might not have been able to top up enough.

Size is also a significant determining factor in your power consumption, particularly for heating and cooling. Despite choosing a tiny house myself at one point, I don't think living in one is suitable for most people. You want to right-size your home for your circumstances.

The sweet spot is enough square footage to live comfortably, but not so much that you're paying for heating and cooling the space you don't use. To determine this, you will need to take into account how many people will be living in your home, as well as the storage and other needs of your space.

When you live off the grid and on a homestead, your home becomes more than just a place to reside; it is also the place where you get a lot of your work done. Instead of buying all your food, you start to grow your own food, cook, preserve, and store it. Instead of buying certain things, you make your own items like candles, soap, and furniture, and you mend broken items. You might homeschool your children, run a side business or work from home, sew your own clothes, process a deer from a hunt, or dry herbs or medicinals.

All these things that the average person has externalized by buying them at the store or outsourcing the labor to an outside enterprise suddenly need to be done in your home. That inherently takes space. Some can be done outside, in an outbuilding, or in some flexible space in the home, but it all needs to be done somewhere.

This is one of the more difficult decisions when building a home: right-sizing it for your needs. The best way I've been able to think about this is to consider your current living space, how it supports your needs now, and what would change in an off-grid homestead scenario. The challenge can be if your current home isn't laid out well or lacks storage.

Before moving into my tiny house, I lived in an apartment that was about 1,000 square feet. When I moved into my tiny house, which was 150 square feet plus a sleeping loft, I found it to be much more comfortable than my apartment because it was designed for my specific needs.

The apartment I used to live in lacked storage, was awkwardly laid out, and had many hallways that wasted a lot of space. My tiny home had tons of storage designed around what I needed to store, was carefully laid out, and had no hallways. My apartment's only advantage was a home office, which my tiny house didn't have.

I considered how it felt to live in all these different spaces and asked, "Was this a lack-of-space problem or just a bad-design problem?" From there, I walked through each room and wrote down what I wanted in terms of space and design. Then I totaled up all the rooms to get my final floor plan size, adding in space for hallways and a little buffer.

My list included a home office, a large pantry, a bigger kitchen with plenty of room for canned storage, etc. That landed me at 1,650 square feet, and I found a design that was 1,750 square feet with minimal hallways and checked all my boxes.

By figuring out the right amount of space, you limit the need to pay to build and climate-control space you don't need. That, in its own right, will be a huge benefit when building your own off-grid home because it limits the power used to heat and cool it.

Alongside choosing the right size and climate, you need to select the right HVAC system. Most homes today have a very low efficiency rating. Efficiency is measured in SEER (Seasonal Efficiency Energy Ratio) ratings, and a lot of technicalities go into assigning this rate. At the time of writing this, the average rating is around 14, with the highest (best) rating in the 20s.

In my experience, if you want to cool your home while also using solar energy, you'll want something with a SEER rating of 20-plus for it to be practical. It will still be very expensive to power a system of this kind, but it is possible. Units with a SEER rating of around 14 are simply too power-hungry to be affordable.

Another big consideration is whether the HVAC unit is single-stage, multistage, or variable. Your most affordable units will have one or two stages, which for a single stage means it's either on or off. The downside to this is that when it's on, the unit is drawing a lot of power, and that power level stays high until the unit shuts off. Two-stage is similar, but it can pull a medium or high amount when on.

Variable speed means that the unit only uses the power it needs for the cooling required. If you just need to maintain an already cool house, it will only draw a small amount of energy. If you come home to a warm home and need to cool it quickly, it can ramp up to cool it fast. The power draw can scale at any increment to the needs and only uses the minimum power necessary.

Higher SEER ratings and variable air units tend to come in the form of mini-splits. The compressor is located on the outside, an air handler is mounted inside, and a coolant line connects the two. They are great units and were what I chose for my own tiny home while 100 percent solar with batteries.

With everything, there is a tradeoff. For mini-splits, the tradeoff is the availability of repair technicians should you ever need one. Fortunately, prices have come down quite a bit, but while these units are pretty affordable, labor for repairs is the catch.

I've never encountered an HVAC tech who said they couldn't fix one, but I find that they understand the units in theory more than in practice. They have experience with normal HVAC systems and are pretty handy, but in the end, they tend to replace units rather than repair them.

Parts can also be challenging to source. Traditional systems have many standardized parts, whereas mini-splits tend to have many custom parts that aren't regularly stocked. In finding parts for my own system, I discovered that after about 10 years it was nearly impossible to get the circuit boards that control these units.

That said, mini-splits are going to become increasingly common, and technicians will therefore have gained more experience with them. They are also the only real option if you want to be totally off the grid in a hot climate, unless you want to always be hot. For a short while, that may be fine, but longer term, it isn't practical and will push you to abandon the pursuit of living off the grid.

Fans can only get you so far. They're great to have for edge seasons or to rely less on a mini-split, but they're not a replacement. Likewise, with heating, keep in mind that as you age, you might not be able to split wood yourself or haul it in like you once did. Having a button you can press to heat or cool your space will be key to your ability to age in place.

The other part of heating and cooling is the insulation of your home. Investing in good insulation will pay dividends on your home's performance and reduce the demands on your power consumption. Insulation today is also very affordable, so doubling your insulation can really move the needle.

There is a point of diminishing returns, but at the very least, you want to use a 2"× 6" framed outside wall insulated with R19 insulation, and if you can, use an outer insulation of one inch to minimize thermal bridging. A really great solution for this outer insulation is one of the newer sheathing products that comes with a closed-cell foam layer already applied (for example, Huber Zip-R panels), meaning that you install it just like you do any regular sheathing in a single step but you're adding a layer of extra insulation.

Increasing your insulation isn't just suitable for heating, either; remember that all insulation does is slow the transfer of heat, so while it keeps heat in during the winter, in the summer, it keeps the heat out. This means that it's universally better to have a well-insulated home, regardless of the climate.

After heating and cooling, the water heater uses about 15 percent of your home's power. Gas-powered tankless water heaters are a pretty simple fix for those off-grid. The challenge with traditional tank water heaters is that they always need power to heat the water, whether you're using it or not.

This always-on power draw means that your off-grid power system is constantly being drained. That's where propane or natural gas comes in. While I don't love the use of fossil fuels, they can offset many of the costs of a solar panel array. If you have a larger hydro-power setup, it may not be required, but for those on solar, these gases are a big help.

A tankless water heater heats the water only as you need it, and the rest of the time it uses very little to no power. What is great about these units is that they are getting more affordable than traditional tanks. They also will keep the hot water coming as long as you need it, never running out like a tank that only holds around fifty gallons and can take a while to recover.

I want to prepare you for all the plumbers, builders, and others who are likely to warn you that tankless units are difficult, expensive, require constant maintenance, or simply don't work well. This mindset may have been correct long ago, but tankless water heaters have now been around for decades. Their detractors are often simply uninformed about the improvements and refinements that have been made in recent years.

Today's units are well built, are easy to use, require minimal maintenance, and are so much better than a tank option. For those who live in areas with hard water, you will need to do a flush once a year, but you can pipe your unit with a ball valve and a system flush hookup on your hot and cold to make this a simple process.

A flushing kit includes two hoses and a small pump, plus a few gallons of vinegar or a bottle of descalers. You let that pump run for 30 minutes while you do something else, and you can flush it to make the inside of your unit as good as new and last just as long as a regular water heater.

The use of a gas-powered tankless unit means that you only need a few watts to fire the igniter and work a fan while it's running. If you choose a unit that is placed inside the home, you don't have to worry about a power draw on an internal heater to prevent it from freezing in cold temperatures.

The only downside to an internal gas tankless water heater is that you'll have to vent it outside. Many plumbers prefer the outdoor-mounted versions because they are simpler to install, which means that they can make more money by getting to the next job more quickly. Sometimes the house design necessitates this, but with some planning, venting can be simple, and your unit will use a lot less power and last longer, too.

After your water heater, you have your washer and dryer. Washers are not too power-intensive, at around 500 watts while on; the saving grace is that you don't run these very often. It can be pretty simple to cover the power consumption with a solar array and batteries, particularly if you prioritize washing clothes when the sun is shining. This lets the power flow from the panels through your system directly to power the washer.

Dryers are a bit trickier. Newer dryers use around 800 watts of power to dry clothing, and older units use up to 1,800 watts. This can be mitigated by switching to a gas-heated dryer or opting for a clothesline.

In general, it's worth upgrading to a modern, more-efficient washer and dryer if you have older units that you'll need to replace in a few years or you're buying new for the house anyway. The power savings are such a leap forward that it will make a significant difference when it comes to the cost of your batteries and panels.

Finally, the next largest draw of power is lighting, which, in a traditional house, uses around 12 percent of the home's power. Just opting for LED lights in all your fixtures will instantly reduce that 12 percent to 3–4 percent. With today's LED bulb options in every shape, color, temperature, and size, there isn't much reason not to do LED lighting.

With all those changes—choosing a good climate, reducing square footage, insulating well, and focusing on the largest power consumers—you put yourself in the best scenario to power your home off the grid. Optimizing these elements will save you thousands of dollars on your solar or similar systems.

One commonly cited off-grid home feature is "passive solar house design." This idea includes several design approaches that utilize the sun's position and interaction with the home to improve its energy performance. This concept has been codified into off-grid home lore to the level of becoming dogmatic, but it is generally an older school of thought that has become outdated.

There was a time when homes in the United States were made poorly, lacking air sealing, utilizing power-guzzling appliances and HVAC, and having minimal insulation. But today, we've learned a lot about how to build better homes and benefit from efficient appliances that are required by regulations.

Passive solar houses focus on orienting the house in a particular direction, extending eves to block the sun in the summer but allow those rays to come in during the winter. They tend to use many windows on the south side of the house to let much of the sun's heat in during the winter, and they make use of deciduous trees whose leaves will block the summer sun but allow the light to stream through in the winter when their leaves are gone.

This is all very clever, but in today's world, the unit economics of building means we can eclipse all these gains for less money and do so more reliably across many climates. Building more-efficient homes without bending over backwards to meet specific design criteria is more practical today and very effective. Beyond just working better, it also gives us more flexibility in our designs.

Instead of building the house to face south, we can orient the home toward our scenic view, align it to the street, or place the home in a way that will reduce site costs. Instead of having a wall of extra-expensive windows, we can allow for good natural light and then double down on good wall insulation, which is three times more insulative.

The ideals of passive solar houses are noble, but in practice I found them to be less effective in most climates and in a world that's getting warmer. Just make sure to consider all the dynamics at play before implementing these approaches.

The last part of building an off-grid home to consider is where you're going to place it in relation to where you'll generate your power. The distance between these two points is a key factor because of the voltage drop you experience over longer runs of wire.

You can find online charts and calculators to figure out the losses you'll incur, but essentially, you lose a certain amount of energy to the resistance of wires. This amount may be very small, but compounded over a few hundred feet of wire, it can add up. There are tricks you can employ, like using larger wires or starting at a higher voltage so that even after the drop, you'll have enough energy where you need it, but nothing beats just being closer if you can manage it.

This may mean that you put your solar panels on your roof, clearing extra space on your land next to your home to ground solar panels, or doing some extra earthwork to get closer to your hydro generator by a stream. The farther you are away from these points of generation, the more it will cost later, and in some scenarios, it won't be possible to bridge the distance at all.

With these considerations in mind, it's time to start the construction process—which begins long before you break ground. We've already touched on the pros and cons of doing this work yourself or hiring it out, but since building your homestead is such a major undertaking, it's worth considering in greater detail. A middle ground between hammering every nail yourself and outsourcing the whole project is to become your own GC (general contractor) and then subcontract out certain parts of the project while doing other parts yourself.

General contractors are essentially project managers with building knowledge. They oversee all aspects of the build, coordinating the entire process, securing permits, finding and scheduling subcontractors for each of the main trades, ensuring that things are done well and that it all passes inspection.

For this they charge a fee, which can be calculated in a few ways, but it often shakes out to about 10–20 percent of the total build cost. A common phrase you'll hear is "cost-plus," which means any cost the builder incurs plus a markup. This typically includes all the materials that they purchase to build the home, the labor that went into assembling those materials, permit fees, taxes, company overhead, and insurance.

For many homesteaders, taking on this role themselves can save 10–20 percent of the entire building cost. Doing so comes with real challenges, but it's possible and a great opportunity to save money if you can successfully run the whole thing yourself.

If you're considering this route, the first thing you'll need to determine is whether your county allows for it. Not all do, but if you're in a more rural area, it's likely possible. The challenge becomes when you want to be your own GC or build the entire thing yourself, and you want to get financing from the bank.

Most banks will require a GC license to be attached to the loan, and while many counties allow for owners to build their own houses, that doesn't give them a license. To become licensed, you usually need to learn a bit about running the business, take a test, and then document that you've worked a certain amount of time under another GC. Each county will be different, but typically you can't just apply, pay the fee, sit for a test, and get the license.

So how do people get this financing and build their own homes? Typically through smaller banks or brokers, but even this approach is going to be difficult, if not near impossible. Banks are all about managing risk, and if you're building a house without a GC license, then you're an unproven builder in their eyes, even if you've worked in the trades for years.

Being your own GC has a lot of merits, but if you're not confident that you can pull it off or you're simply not allowed to, finding a builder is still a great option. While it does add cost, sometimes it's worth the money, particularly if you find a good one.

This was the route I chose when building on my land. While I had experience building my own tiny house, I knew it would be challenging to get a mortgage if I wanted to act as my own GC in my area. My county does allow for owners to be their own GC and even do some of the trades work, but after considering the matter carefully, I still decided to hire someone.

First and foremost, I know how much work really goes into building a home. General contractors earn their money; it's not a small project, and you would be shocked at the number of details and decisions that they need to account for in the entire process.

In my situation, I was building several hours away from where I was living, which meant I would need to either rent a place close by or set up something temporary. Both options come at a cost, which I calculated according to the length of the build time.

I also know how hard it is to get trades people lined up and scheduled. Building in a rural location meant I had fewer people to choose from, but first I had to find them. Finding trades workers, getting quotes, or just getting them to call you back can be a real hassle.

This is particularly true because you're competing for the subcontractors' time with builders who have already established a relationship. If a builder has been good to them in the past, paid them on time, made the work easy, and if the subcontractor does a good job, it will lead to consistent work for years. Your single job will never be a priority unless they have a random gap in their schedule.

If you do find a subcontractor with time to kill, it unfortunately may be for a reason. While there are legitimate reasons a subcontractor could be free—maybe there was a cancellation or they're just starting out—they may also be available because they don't do good work, don't show up on time, or are difficult. Of course, it's possible to find solid professionals who have openings, but with the major labor shortage in the trades here in the United States, this is more the exception than the rule.

The other factor I considered when deciding whether to hire a builder was my own lack of experience. Being a newer builder undoubtedly meant that I would make mistakes, and while I know more about building a home than the average person, I estimated that I'd make a few minor errors, along with at least one large one. Each of those mistakes would cost money, take time, and get me hung up in inspections. I estimated the value of those errors and I realized that hiring a builder meant that mistakes likely wouldn't happen, or if they did, it wouldn't be my financial responsibility to fix.

The consideration that really clinched my decision to hire someone came from accounting for my time. Building a house takes a lot of it—on average, a home takes around 6,000 hours of work to complete when you account for all parts of the process. That assumes that you're hiring experienced workers and those who are specialists in their trade. For someone like me, I don't think it would be unrealistic to say that my lack of contacts in the area, all the travel time, and my sheer inexperience could add a couple thousand hours.

Being a desk worker, I also knew it would take me some time to get used to the physicality of building. It was manageable when I was building my tiny home, but I certainly never had any trouble falling asleep at night, and despite being young at that time, I felt sore every morning after a workday. I can only imagine what that would be like today when I'm fifteen years older and the house is ten times bigger.

With all that in mind, I realized that I'd likely have to step away from my work for a while to complete the build in a reasonable amount of time, or it would drag out for years. The reality was, when I sat down and crunched all the numbers, I could earn more working than what I would have saved by building the house myself, even after I accounted for the cost of hiring a builder.

My last consideration was the amount of time before I could actually get on with living in my new house and focus on building out my homestead. I estimated that it would take about 18 months for me to lead the build (using subcontractors). With a builder I was looking at about 6 months, and in that time, I could be preparing to start projects as soon as I moved in.

All in all, I found that by hiring a builder, I could save money, get on the land a year early, remove some risk from the process, make it easier to get financing, and avoid all the stress and details of the build itself. It was an easy decision. When making your own choice, you'll have to evaluate your particular circumstances and do what's best for you, but hopefully reading about my thought process will help you in your journey.

Once I decided to hire a builder, I had to go about finding one. Luckily, builders tend to be easier to locate and talk to than your typical trades person, but it can still be difficult to find one who happens to build in your area. This was the case for me: I could find them, but they were so busy close to their offices that they wouldn't build farther away than about a twenty-minute drive.

Ultimately, my best resource was the local home builders association. You can search online for your town or county's name plus "builders association" and likely find their website. These association sites often include a list of members and their contact information. I started at the top and called each one of them. If they didn't serve my area or were booked out for a long time, I'd ask if they knew anyone to recommend.

This ended up yielding the best leads. I just kept calling and asking for referrals and in the end I had ten calls with builders. I narrowed it down to five builders to come out to the land, then I had three give me a bid for the job from the same specifications and floor plans.

When talking with potential builders, I considered how they presented themselves, how they described their process, how willing they were to answer my questions, and whether they showed up on time or called me back promptly. In the end, I made my selection of the final three because they seemed to have their affairs in order, they followed a fleshed-out process that they could explain well, and we had some level of rapport.

Finding a builder took a while. Even though I made a consistent effort, it took the better part of a year to make the calls, chase down new contacts, set up meetings, drive out to the land, and go over all the details of their bids. Going through their proposals was relatively simple, because the bids ended up coming in at very similar price points, mainly because I had laid out specific choices in the build and materials I wanted, and had already decided on the floor plan.

The challenge when determining why a builder might charge more or less than another often boils down to three main things: their business overhead, their fee structure, and their material selection standards. Builders with nice offices, large teams, nice trucks, and lots of marketing means that they need to pay for all that somehow. I found that the builders who operated a small team were often a pretty good value for the money. They had some overhead, but not a huge amount, and they had the necessary experience and were able to keep their subcontractors busy.

You can compare builders' fees pretty easily. Just be wary if they aren't willing to talk through how they assess their fees. A good builder will be up-front about their fee and happy to talk you through it.

The last aspect of the build decision is material selection, which can be challenging. Builders default to certain materials for each part of the build unless you specify otherwise. There are many different brands of materials, and within each brand, different tiers or grades of product lines.

You can't ask the builder the details about everything because there is so much that goes into constructing a house, but you can ask about the largest line items and then compare across the builders you are considering.

Your biggest material line items will be
• Doors and windows
• HVAC units
• Roofing material
• Siding material
• Sheathing material
• Cabinets: carcass material, slides, etc.
• Foundation type: slab, crawl space, basement
• Foundation method: poured or block
• Driveway type and amount
• Tile work

Get your quote from the builder, then set up a meeting after you've had some time to review it. Let them know that you want to talk through it and that you need to understand what assumptions they baked into the prices of the items in the list above, along with other questions you have.

In that call, you can ask what brand they typically use for each item, why they made that choice, how that brand and product tier compares to other options, etc. Take notes on all of this and you'll start to find that certain builders skew to more budget materials, some choose the best of each category, while others might be somewhere in the middle.

Asking why specific options were selected is a really great way to understand your builder better. You get to see how well they explain things, what they value when making decisions, and how organized they are. If a builder doesn't lean into your discussion or seems to hold back, it might be a red flag. If they can't communicate well, you'll get an idea of what working together would be like. If they give nuanced answers, easily explaining what they choose and why they choose it, you might have a good partner for your home build.

After you do this with all your builders, you can start to compare across all the materials a little better. That, combined with their fees and a sense of their overhead, will let you better understand the final price in the quotes. While each of these aspects is important, ultimately you want to find someone who gives you confidence about the project and respects your priorities. You don't need to belabor your meetings, but a home build is a large project and warrants careful attention.

Once you've decided on a builder, you want to have a few items in place before signing anything or handing over any money. First a set of floorplans: they don't need to be 100 percent finalized but should be very close. Next is the budget, laying out all the line items for the build and clearly showing prices and fees. And finally, your contract: you should always have a legal agreement that lays out each party's responsibilities, a timeline, and other important elements.

With these three things thing in hand, you can approach the bank and discuss financing options. Once you are pre-approved and the underwriter has reviewed your financials and the builder's documents, you can sign with the builder and the bank. Soon, it will be time to break ground.

STEP 37   POWER YOUR HOMESTEAD

—

Powering your off-grid home involves a lot of factors, so I want you to think of this part of the book as a way to navigate the big decisions. However, we simply can't cover every last detail. The reason is twofold: first, multiple books could be written on this topic alone. Second, if we got too far into the details, this book would become out of date very quickly.

I want to focus on making more-significant design decisions and help you understand what questions to ask next. There is a pretty steep learning curve with all of this unless you happen to be an experienced electrician or engineer. For those without that expertise, this guide can make you a more informed shopper when considering whom to hire to do the work, should you decide to loop in a contractor.

First, let me start with wind and hydro power generation because they require very specific circumstances to be viable. For wind, the best advice I can give is this: unless you look around and see wind turbines being used in your local area, it's most likely not going to be practical for your location as your primary power source.

Even if there are wind turbines in your area, using wind as a power source can be challenging. Turbines have many moving parts that require annual maintenance, part replacement, and sometimes repairs. The other challenge is that most of the advantage of wind turbines in a good wind area comes from the turbine blade size.

The larger the turbine blade, the better, which is obvious, but if you were to map the power curve in relation to the size of the turbine, it's not linear; it's a logistic growth curve, which is shaped like an "S." What that means is until you get to a very large blade, you don't make a lot of power. There is a point at which you hit critical size and the power production climbs almost vertically until it levels back out when it hits a plateau of sorts.

This presents a challenge to smaller, home-scaled wind power generation. Typically, you can't afford to get a turbine big enough to make enough power. Even if you live in an area without an HOA, state or national codes will likely have some say because of the height required by wind turbines. Then you have to factor in the continual cost and effort of maintenance.

There are several types of wind turbines: horizontal and vertical versions, roof ridge turbines, and many more. While they all have their pros and cons, some with distinct advantages, you still can't get away from the laws of physics that govern them all.

When it comes down to it, you usually can buy the equivalent production in solar panels for about half the cost, and solar panels don't require much maintenance other than giving the panels a good rinse once a year. The only real advantage that wind turbines have over solar is that the wind blows at night and when it's cloudy, times that solar doesn't work well. Wind might be an excellent secondary option in that regard, but you're likely better off sinking those additional dollars into more batteries.

The next option for power generation is hydropower. If you have a scenario that makes hydro possible, you've lucked into an incredible power solution. This approach requires a special set of conditions to work well, but when it does, it's impressively consistent, relatively low-cost, and can make a lot of power 24/7.

The two biggest factors you need to look for are head pressure and flow rate. Head pressure is how many PSI (pounds per square inch) you can generate from a downhill slope. Flow rate is how many gallons per minute your water source has and what you can realistically divert in order to generate power.

What this means practically is that you need your water source to start up high and then flow downhill to where you can put your turbine. This drop needs to be at least 10 feet, but more is better. Water will naturally flow downhill; we just need to make sure we can keep our pipes at a constant download slope as we run the water.

Flow is measured in gallons per minute. A small trickling brook will likely not have enough water flowing through it, whereas a consistent stream or small river will have quite a bit of flow. It's important to make sure that your water source is consistent year-round. The flow rate will change over the course of the year and in relation to rainfall, so use the worst flow rate in all your calculations.

Certain scenarios allow to dam your stream to create a pond; this can help mitigate changes in water flow by capturing more water when it rains and doling it out slowly in the dryer months. All this work—creating ponds, diverting waterways, and installing hydro equipment—typically requires government approval because it impacts watersheds and natural ecosystems. Many townships have review panels to evaluate these types of changes; the factor you'll have on your side is that you're diverting only a portion of the water and returning all of it back to the original stream.

To get a rough idea of what your water source could generate, this formula can serve as a quick back-of-the-napkin calculation:

$$\text{Dynamic head (ft)} \times \text{Flow rate (gpm)} \times 0.18 \times \text{Turbine efficiency} \\ (0.40) = \text{Power output (watts)}$$

You can see here how important a good height advantage for your head pressure and good flow can be in influencing how much power you can generate. If you're operating at the minimum head pressure of a ten-foot drop, you're going to need a lot of flow to compensate.

With all this in mind, if you are looking for land, you can add this to your shopping checklist / wish list. Just know that any land with a good water source is expensive. Everyone wants it, even if they don't use it for hydropower. But on the other hand, if you could find the right water source, you would never have to pay a power bill again.

I've seen three things that can derail a hydro plant: seasonality of water sources, freezing temperatures, and heavy rains leading to flooding. Every water source ebbs and flows with the amount of rain that falls throughout the year, and if it slows down too much, you might not have enough power.

In places where it gets really cold, you'll have to contend with your water freezing, which means you'll need to make sure you're clearing away ice from inlets and strainers and protecting your turbine from freezing solid. You might think about bringing your turbine inside an insulated shed that has a heater in it, and you'll have to go out each day to clear ice from your intakes.

Finally, a lot of rain falling all at once can lead to surges in water and result in flooding. This isn't much of an issue for the turbine because only so much water can be carried by your pipes. What could become a problem is floating debris damaging your pipes, spillways, and inlets. You can have all your pipes washed away in a flood, but you can plan for this and install things to have most of your pipes routed away from the flood zone (while still keeping a constant downward slope) and keeping replacement parts on hand.

Now let's talk about solar power, which I'm going to spend the most time on because, for most people, it's the most practical approach that works in most areas and has so many benefits. Again, I'm going to keep this high-level because solar has been progressing steadily for decades, and if I were to start naming products, exact numbers, etc., most of the information, beyond basic physics and electrical concepts, would be out of date in about six months.

Solar panels have come a long way, and the price per installed watt has dropped steadily over the years. Solar has benefited a lot from mass production, improvements in the panels, and lower costs as things have scaled up.

Solar today is a competitive space where you can get great panels for a reasonable price that will likely last you for at least 20 years, if not longer. I've found that the panels last so long that when they really start to wane in performance, the technology has improved so much that it's often worth upgrading at that point anyway.

When I first set up my solar array, the batteries I used were lead–acid batteries, and they served me well for about 10 years. I chose the lead–acid batteries because they had the best capacity-to-cost ratio at the time. Lithium-ion batteries were available and great options, but for a similar storage capacity at that time, they cost about three times the price of lead–acid batteries.

Jumping forward to today, that cost has dropped from about three times the comparable price of lead–acid to about 1.5–2 times the price. Depending on deals you can find and tax credits available, you can sometimes narrow that gap even further.

Today, I might spring for the lithium-ion batteries because, while they may cost a bit more, they also are less work to maintain and have better performance. By the time I change out my batteries in the next 10-ish years, it will likely be cheaper to go with lithium-ion because, as the economy of scale continues to improve, the cost will be pushed downward even further, particularly with increasing demand for electric vehicles.

I paint this picture because I want you to understand how fast these things are improving and to know that if you're reading something over a year old about solar technology, it's likely representing outdated thinking. This is a challenge when you ask for advice from those who've done it before, because their context was frozen in time when they built their system, so what is true for them, won't necessarily be true for you.

With that caveat out of the way, let's start with the basics. At its core, solar energy is about capturing the sun's energy in a panel and converting it to a useful form of electrical power. That power can be stored in a battery for later use, or it can be used right away on something that needs power at that moment.

A solar system has several key components to make all this happen. You have solar panels, which collect the power, a converter to manage the power flow, batteries for storage, and an inverter, which turns DC power into AC power.

Before we get into all the specific parts, we first need to know what size system we are building toward. To understand that, it's vital to know a few formulas. Please stick with me here. I've tried to simplify this as best I can, but it does require some math.

The first formula concerns how watts, amps, and volts work together. You take amperes (generally called "amps") and multiply them by volts to get your watts in a formula that looks like this:

$$\text{amps} \times \text{volts} = \text{watts}$$

If you remember your days in algebra, you'll know that you can rearrange this formula depending on your information and what you need to figure out. Here is the same formula rearranged based on what you need to solve for:

The original formula: amps × volts = watts
To solve for volts: volts = watts ÷ amps
To solve for amps: amps = watts ÷ volts

Other aspects of math will be watts versus kilowatts, which are abbreviated as kW. These are the same, but the kilo stands for 1,000, so if you divide your watts by 1,000, you'll get your kilowatt number. The opposite is true, too: if you have a number expressed in kWs, you just multiply it by 1,000 to get the wattage.

The final thing you should know about formulas is that you'll sometimes see a unit expressed with "hours" in the unit's name. You'll see this in things like kWh, which stands for kilowatt-hour or amp-hour (Ah). When figuring out the size of your battery bank, these types of units are all about how much energy (power, measured in watts, or current, measured in amps) you need multiplied by the number of hours you need that energy for.

If you use 1,000 watts and need that amount of energy for five hours, you'll multiply those two numbers together and combine the unit labels. In this example, you'd have 5,000 watt-hours, where you have combined the watts from the first number and the hours from the second number.

Another example that might be helpful is amp-hours. In this case, it is the same thing; if you need 50 amps for 5 hours, that will be 250 amp hours.

The question you might ask yourself is, What about volts? Without getting too technical, volts are the "pressure" that an electrical system can exert. Instead of being a measure of how much "work" can be done, like amps and watts, volts are technically a "potential" rather than "work." All that means is you won't see or need "volt-hours" because they aren't a thing.

Typically, you'll know the voltage in your system because it's either coming from your panels, indicated on the panels' spec sheet, or what you've decided for your system's battery voltage (12 volts, 24 volts, or 48 volts), which we'll get into soon.

Now that the basic math and electrical theory are out of the way, we can calculate how much energy you use and what that means for sizing your system. The simplest way to do this is to write down all the power-consuming items you use, calculate their wattage consumption, and consider the number of hours you use each item per day.

The easiest way to figure out the watts that something uses is to get a so-called Kill A Watt meter. You plug the item into the meter, and it will tell you how many watts, volts, and amps the item is drawing. In a pinch, you can look at the item's label and write down the watts indicated or calculate it by multiplying the amps times the volts to get the watts. This method usually overestimates, but it will be close enough.

For each item on your list, multiply its watts by the number of hours you use it per day. Add up all those numbers to get the total power usage per day. With that number, I usually add a 20 percent margin for system inefficiencies. The total gives you how much power you need to produce in a day.

From there, you'll need to figure out how many "solar hours" your location has per day. This will differ from the total hours of sunshine, because your solar panels will only work at peak efficiency when the sun is overhead and the panels get direct sunlight.

At my location, I get about 6 solar hours in the summer, and in the winter, because of the shorter days, I get only 3.1 hours in January. With that in mind, I built my system planning off the worst day of the year, which was 3.1 hours for me. Your location will have different numbers, and you can look those up online with a simple search.

With those two numbers, the total power you use in a day and the solar hours, you can figure out how big a system you need. You'll divide the total watts required by the solar hours. So, in my case, I needed 4,000 watts per day, divided by 3 solar hours, which gave me a system of around 1,300 watts.

Now, if it was cloudy, I wanted to have about three days of power stored in my batteries so I took that estimate of 1,300 watts and multiplied it by 3 to give me 3,900 watts. I then rounded up to 4,000 watts of solar panel output for good measure. That helped me figure out how many panels I'd need by dividing that number by the number of watts each panel could produce.

In my case, my panels were 265 watts, so 4,000 divided by 265 meant I needed 15 solar panels to meet my power requirements. You'll have to figure out your numbers based on your usage and your local solar hour production.

Now let's talk about the solar panels themselves. Solar panels, as mentioned before, convert the sun's energy into electrical current in the form of DC (direct current) energy. There are a lot of details about the type of panels, the technology behind them, the density of cells, and different wattages, but if you keep two things in mind, you can quickly navigate all those decisions.

First, you should buy a good-quality panel from a manufacturer with a solid track record. There are two schools of thought on what constitutes "good quality": those listed as "tier 1," and brand reputation from those in the field.

Tier 1 solar panel manufacturers are generally going to be a pretty safe bet when it comes to choosing the brands, but it's important to note that this list is actually evaluated on the financials of the company, not on any solar testing done. Nevertheless, many feel this is a good measure because those companies that are fiscally sound tend to be that way because they offer good products and are more likely to honor their warranties over the long term.

This school of thought isn't without its detractors. They'll point out that certain industry leaders aren't on that list and also that the panels have not undergone any actual technical testing.

I've found reputation helpful when you talk with a local solar installer in your local area. If you ask what brands they currently like to use, you'll get a good list of quality solar panels. This is because these installers are always looking for a good deal, but they don't like to cheap out because they know they'll have to fix it if the panels don't work well. Simply put, they tend to choose a good mix of quality and cost so they can do the job once and move on.

The other main thing you need to consider is the panel voltage. You want the output voltage of your solar panel to be larger than the voltage of your batteries. So, if you're doing a 12-volt battery system, your panel voltage should be more than 12 volts.

Those two things will steer you in the right direction almost every time. You'll want to avoid direct-to-consumer brands like Renogy, Jackery, Goal Zero, Bluetti, and anything you can find at your local hardware or farm supply stores. These brands are good hobby options, but they come at a premium, and the ones you find at a local store are usually very low-quality.

Instead, buy your panels from a dedicated solar supplier or local installer because they get professional-grade panels by the pallet directly from the manufacturer. These tend to be the best manufacturers at competitive prices from dealers who can sell with full warranties.

Once you've chosen your panels, you'll need to decide how to wire and arrange them. This involves wiring panels in series or parallel. When you wire solar panels in a series, their output voltages are added together while their output current (amperage) remains the same. Conversely, solar panels that are wired in parallel have their output increased in amps by adding them together, while their output voltage will be the same.

Your individual scenario dictates whether you should wire your panels in series or parallel. You can also do both, wiring a few panels in series to raise the voltage and then wiring those groups in parallel to raise the amperage.

So how do you know which is right for you?

Figure out which charge controller you are going to use—we'll talk about that next—and follow the specifications indicated on its specification sheet. You can typically also find that information right on the charge controller.

From there, find the battery bank voltage that the charge controller can handle: 12, 24, 36, or 48 volts. That specification sheet will also tell you the maximum "PV input voltage," meaning how many volts are coming from your solar array, known as the "open circuit voltage."

Finally, the specification sheet will indicate the max "PV input wattage," which is the total number of watts coming from your solar array, depending on your battery voltage. With those three numbers, you'll have everything you need to calculate different scenarios and determine the best wiring approach.

When you calculate different combinations of series and parallel wiring, you want to get as close as possible to the maximum rating of your charge controller. This is where someone with more experience can really come in handy. They have done this enough to be able to quickly intuit the right combination, while someone without much experience will have to game out each combination.

The one curve ball you need to consider is when you have longer runs of wire between your solar array and your charge controller. This, again, is because of voltage drop. In such a case, it's better to opt for a higher voltage from your panels (by leveraging wiring in series) so that even when the voltage drops over the run of the wire, you still have enough voltage to be useable at your charge controller.

Next up are your converters, typically referred to as charge controllers. These take the power from all your panels, combine them into a single stream of power, and then handle charging your batteries at precise amount of volts and amps to not damage them and keep them at peak performance.

Choose a good name brand that is an MPPT-style controller. (MPPT stands for "maximum power point tracking.") Other options like shunt controllers or PWM (power width modulated) controllers exist, but they are an older technology; if you find a charge controller that uses them, it tends to be old or made by a lower-quality brand that should be avoided.

So first make sure to only consider MPPTs, because they typically are 30 percent more efficient and they can handle a higher-voltage array, which is necessary when you're trying to power an entire home. Here again, choose quality brands by getting advice from local installers.

It's worth noting that your array will likely be larger than what a single charge controller can handle. Choose one that is made to handle larger voltage loads, but realize that you'll likely need more than one of them. Since you'll have more than one, make sure the charge controller you choose can communicate with multiple charge controllers to manage your larger system. It's best to buy all your equipment at the same time once you've finalized all the details so you can get matching charge controllers. I also suggest that you stick within the same brand for your inverter as well.

Your charge controllers will handle all the power coming in from your solar array and manage the charging of your batteries. They will also make sure the batteries are not overcharged and will turn off the system if your batteries get too low, in order to prevent damage.

After the charge controller come the batteries themselves. The two big decisions you'll need to make are what type of battery you want to use and what voltage system you want to use. For a long time, most systems were set up as 12-volt systems, but today's systems tend to be either 24-volt or 48-volt, with the latter being more and more common for homes.

The higher the voltage, the more efficient your system generally is, which means you can save a fair bit of cost on things like wires, etc. The only real reason not to go with 48-volt is when you can only get an inverter that can only handle 12-volt battery banks or 24-volt battery banks. In general, though, a high-quality inverter these days will be able to handle them all, so go with 48 volts to power your home.

Now, the type of battery you choose will typically be driven by cost more than anything else. Depending on when you're reading this, battery technology may have progressed to the point where some of the newer types are more affordable.

Right now, you'll find a few basic options that will likely be common for a while; the prices will just change over time. The first type is lead–acid batteries, which has been the standard for a long time now because they work well, are well supported by your equipment manufacturers, and are cost-effective. Lead–acid comes in a couple of subtypes: flooded and sealed.

Flooded lead–acid batteries are the most common and require some maintenance over time. That maintenance typically means keeping the cells topped up with distilled water, running an equalize setting on your charge controller, and keeping the terminals free from corrosion. None of that is difficult, but what I found in practice is that it's more a hassle than hard to do. Life gets busy, and it takes time, but you have to take care of your batteries to get a good life out of them.

Sealed lead–acid is less common, but it comes in the form of AGM (absorbent glass mat) or gel batteries, which are available, but I haven't seen many people install them. They are typically a bit more expensive but do not require the same maintenance, since they are sealed. All you need to do is house them properly and ensure their terminals stay clean. The only downside is that they cost the most, but the convenience is typically worth it unless you can jump up to the next options.

Lithium-ion batteries come in a couple of different types, mainly distinguished by the material they use for the positive electrode. There are many different formulations these days, but lithium iron phosphate is the kind preferred by most now and likely in the future until we find the next great leap in battery technology.

You will see these batteries referred to as lithium iron phosphate, LFP, or $LiFePO_4$, which are all the same thing. They work very well for this purpose, and their performance exceeds lead–acid in almost every regard, except cost. Just make sure to house the batteries in something that prevents them from being exposed to temperatures below freezing. But other than cost, $LiFePO_4$ is better in almost every way.

You will hear some mention of the risk of fires with these types of batteries, but that is largely due to older formulations of lithium-ion. If you're using quality equipment in your system, every single component monitors for the conditions that would cause a fire and shuts them down well before they become dangerous, which would only happen if the battery were to be punctured or severely damaged—not really a concern when they're kept sheltered. This fear of lithium-ion batteries posing a fire risk tends to be passed around online as a solar battery boogeyman, but it has been totally addressed a very long time ago.

So your choice will be between lead–acid and LiFePO$_4$, the latter being a clear choice if your budget allows it. The next question: How many batteries do you need for your system? This will depend on how much power you use and which battery type you select.

Lead–acid batteries should only be discharged down to about 50 percent, while LiFePO$_4$ can be discharged 80 percent (20 percent remaining) before you start to impact the performance long-term. Technically, they can be discharged even more, but you should set these amounts in your charge controller to get the maximum life out of your batteries.

We figured out how much power we needed earlier when we listed all the items we power and how long we used them for. We want to use that initial number, not the one we later divided by solar hours. Multiply the wattage of each item by the hours used, and then add up all your items. That will give you a total in the unit of "watt-hours." Then multiply that number by either 2 (for lead–acid's 50 percent depth of discharge) or 1.2 (for LiFePO$_4$'s 80 percent depth of discharge), which adds a buffer so that you don't have to run your batteries too low.

From there, I suggest adding 5 percent to account for system inefficiencies, so multiply your number by 1.05 to add an additional buffer. That will give you your final battery size in the unit of "watt-hours," which you can use to size your batteries.

This formula is calculated as:

Power requirements × 1.2 (LiFePO$_4$'s 80% depth of discharge) × 1.05 (inefficiencies) = Total watt-hours

Example: 4,000 watt-hours × 1.2 × 1.05 = 5,040 watt-hours

Sometimes you might need another number to size your batteries, specifically "amp-hours." Batteries have an amp-hour rating and a voltage. To get this number, you can divide the voltage of your system (12 volts, 24 volts, or 48 volts) by your number of watt-hours to get your amp-hours.

Continuing the previous example, if your system were to be a 48-volt system:

5,040 watt-hours ÷ 48 volts = 105 amp-hours at 48 volts

Typically, when you're shopping for your batteries, you'll include the voltage and the amp-hours they can supply. For example, if you're looking at a $LiFePO_4$ battery that is rated at 48 volts and 100 amp-hours, you'll note from our example above that we need 105 amp-hours, so that wouldn't be enough. In that case, I'd either choose another battery rated above the 105 amp-hours or get two of the 100 amp-hour batteries.

It's worth noting that I'd always round up rather than down on battery storage. You'll also want to build in extra storage for when it is cloudy or rainy; for me, I wanted three days of "autonomy" in my system, so I multiplied what I needed for amp-hours. Continuing the example above, if I wanted three days of battery power just in case, I'd multiply 105 amp hours by 3 for a total of 315 amp hours at 48 volts, and buy batteries to suit.

With this information, you'll be able to select the battery size and voltage you need for your system. Note that sometimes you need to choose a lower-voltage battery that you'll wire in series to stack the voltage. For example, if you want a 48-volt battery system but the batteries you want only come in 12-volt versions, you'll wire four of them in series to get to your target of 48 volts.

If your batteries only have a lower amp-hour rating than what you need, you'll wire them in parallel to stack the amp-hours. For example, if you want 300 amp-hours of power, but the battery you want to use only comes in 100 amp-hour sizes, you'll wire three of them in parallel to reach your target of 300 amp-hours.

You can also do both at the same time if you need to stack your voltage and stack your amp-hours. Remember that wiring in series adds voltage together but not amps; wiring in parallel adds amps but not voltage. A combo of wiring some in series and then some sets in parallel will give you what you need. This is where someone with experience can intuit the right combination; you'll have to game out the different combinations to get the right mix.

Lastly, we have inverters, which take the energy up to this point, which is DC (direct current), and convert it into AC (alternating current). Solar panels only make power by DC, and your home and everything in it uses AC, so we have to make this conversion before the power is usable.

Many people have seen online discussions suggesting that DC is a more efficient way to power things. Typically, everything in a conventional house is set up for AC. Still, if you're installing solar and constructing your own home, this question becomes pertinent when starting from the ground up. Since solar panels generate DC power, you need to decide how to manage it.

Much of the guidance on wiring a house for DC power originates from outdated sources; these might be old articles on the subject (consider anything older than a year outdated, given how rapidly solar technology advances) or from individuals not up-to-date with the latest equipment.

Historically, the motivation to wire a house for DC power stemmed from two main reasons: there was significant power loss through inefficient inverters (which convert DC to AC) and the fact that, theoretically, DC is indeed more efficient.

In modern times, this rationale falls short because inverters have significantly improved. While there is still some power loss during the conversion from AC to DC, it's minimal. Another point is that any inefficiencies (both from the AC conversion and the inherently less efficient nature of AC) can be easily offset by adding one or two extra panels to your solar array.

This approach becomes even more practical today, because wiring for DC limits you to DC-powered appliances, which generally cost two to three times more than their AC counterparts. This means that even after accounting for conversion losses, you can achieve more power with AC for less money. The savings from opting for AC appliances over DC will leave you with more cash, even after purchasing one or two additional panels.

Additionally, working with AC power in the home is much easier if you need to hire an electrician or get inspected for permits, etc. Both electricians and inspectors tend to be more experienced and comfortable working with AC, which might also be required in your area anyway. In simple terms: convert to AC, add a few more panels to your array, and stop worrying about AC versus DC.

When it comes to selecting an inverter, you'll want to go with a high-quality brand that is a pure sine wave inverter. All the nicer inverters have this, and if you see an inverter with older tech having a "modified wave," avoid that because it can be rough on fragile electronics.

You'll size your inverter by figuring out the maximum wattage you might draw at any given time, so list out all the appliances you could run at once and then add a 25 percent buffer. This will tell you the total wattage your inverter will need to support. For example, if we need to run 3,000 watts at any given time, adding in a 25 percent buffer, you'll want an inverter that is rated for 3,750 running watts.

When in doubt, round up. This gives you a bit more wiggle room and will future-proof for home for anything new you might power down the line. One additional thing to consider is surge capacity. While your normal running loads need to be considered, certain items draw a lot of extra power for a fraction of a second when first starting up.

Today, the only things we have to consider are your HVAC and power tools. A few things, like refrigerators, used to have big surges but no longer do with modern versions. Typically, surges coming from these units are three times their running watts, volts, and amps for about a third of a second, sometimes up to two seconds. Good inverters can handle these spikes without issue, but it's something to check. Take your biggest load, multiply it by three, and check that number against whatever your appliance lists as its surge rating.

Another big consideration is whether your inverter only operates at 120 volts or can also operate at 220/240 volts. You'll want that ability if you're running HVAC, electric stovetops, welders, laundry units, or other 220/240 volt appliances. I mention this because many inverters do not operate at 220/240 volts, so be sure to select a model that meets your needs.

Along with that, you'll want to consider an inverter that can support the battery bank voltage that you're building. Most good-quality ones sized for fully off-grid homes can handle any of them. You just must select it in the inverter's settings menu, but it's something to make sure you check when purchasing.

Another feature you might consider is generator support. Several inverters detect when your batteries are low or when your system draws a very large amount of power, and then automatically starts your generator. This is great when your batteries are low; it will start the generator to charge them up to full capacity without you having to do anything.

If you're going to be totally off the grid, my suggestion is that you build a system with a propane or natural gas generator setup that will work with your generator from day one. If you've sized your system well, it will barely ever run, but once in a while, when the weather is bad for a long time, you'll be thankful for it.

I learned this the hard way when, one winter, it rained for two weeks straight. It was very overcast the whole time, and my batteries bottomed out on the fifth day. Winter is particularly challenging because you need a lot of power for heating, and you tend to spend more time indoors running lights, watching movies, and running your heat. This, combined with shorter days in the winter, makes it hard to keep the power going.

That leads me to my biggest recommendation: build a system for your worst day, and then every other day will be easy. Adding generator support when you need it will bridge any gap and is worth the extra investment.

The last note I'll add here is that if you work with companies that put solar panels on your house to lower your power bill, be aware that they often use inverters that can't run your home if the grid goes down and that don't typically support battery banks. They do this as a cost-saving measure, and I've seen several people get caught not realizing that their inverter can't support those two scenarios. Be sure to ask if you work with a company like that.

In a nutshell, that is what you need to know about setting up a solar-powered system for your off-gird home. You can see that a lot of considerations are involved, so don't hesitate to involve a company that does solar installations. I find that their pricing is usually better than what you can get on your own by buying individual components of the system. Plus, you don't have to worry about shipping costs and import fees. While you're paying someone for their help, the reduced material cost tends to offset the fee, and you will also gain the peace of mind that comes from an working with an expert.

STEP 38   USE THE GRID AS YOUR BATTERY

While we focused on batteries in our last step, I did want to suggest one alternative strategy that works well for many people: using the grid as your "battery" instead of relying on a huge battery bank or even any battery bank at all. This approach involves connecting your home to the grid with a "net meter," which charges you for the power you use from the grid and subtracts from your bill the power you feed back into the grid from your solar panels when you do not need grid power.

This essentially is power arbitrage, and it works so well because, in many states, the law requires a power company to buy your power at a preferred green energy power rate if you're under a certain size.

In my area, normal power costs around 11 cents per kilowatt, but the green energy rate is 21 cents per kilowatt. That means I can sell my power at a higher rate when I don't need it, which adds a credit to my account, and when I need extra power, I can buy it at a lower rate.

In this way, I use the grid like batteries. In many places, any extra money you earn is credited to your account, but typically, it doesn't roll over year to year, so at some point it gets cleared. In some places, the power company will write you a check for the extra amount each month, which is a nice turn of events.

The even nicer part of this is that you can still have batteries if you want a little extra insurance, or you can have a local generator for when the grid goes down. Just talk with your local power company and discuss their policies before you make plans.

STEP 39   SOURCE CLEAN WATER

—

Water is life, and that goes double for an off-grid homestead. Besides drinking, you'll need it for cooking, cleaning, watering the garden, filling animal water troughs, and many other things.

Having lived in a dry cabin for a while and watched many friends try to make it without a good water source, I can tell you that nothing that will push you to abandon your living arrangement as fast as not having good water right out of the tap.

As mentioned in step 17, getting your water source squared away is one of the first things you should do on any new land in order to ensure it will work long-term. There are three main ways to develop a water source: drilling a well, harnessing a natural spring, and capturing rainwater.

First, let's consider how much water you will need. The average American uses about 80–100 gallons of water per day in their daily life. When you're going off the grid, it's good to be conscious of the water you use, but don't be unrealistic. How much water do you really need? Are you going to get by using 10 gallons a day? It's not likely, unless you're already doing that now and it works for you.

To get an idea of how much water you'll need per day, let's break it down into a few categories:

- Drinking water
- Cleaning
- Hygiene
- Watering Gardens
- Animal Needs

For drinking, plan on about a gallon per person per day. You might want to budget a bit more if you're going to be doing a lot of hard labor and need extra hydration. Doing the dishes can take up to about 20 gallons per meal if you hand-wash or 5 if you run the dishwasher. Washing your clothes requires about 20 gallons of water for each load you run. Then you'll want to factor in about five more gallons for miscellaneous hand-washing, wiping down counters, rinsing produce, and cooking.

For hygiene needs, you need to figure 1.5 gallons of water per toilet flush, accounting for the number of people in your household. You can use a composting toilet that doesn't need any water, but it's good to build your system to support a flush toilet just in case you change your mind later.

When it comes to bathing, different people need different amounts of time, but you'll want to factor in 1.5–2.5 gallons per minute based on your shower head. This can really add up; a 10-minute shower requires 25 gallons per person per day, assuming you shower daily. Factor in water for shaving, brushing your teeth, washing your face and hands. All that adds up.

If you plan to garden on your homestead, your needs will vary, but typically, you'll want to give the garden a good soak a few times a week, which can work out to about 500 gallons per 1,000 square feet of garden space. You can conserve water with drip tape run under your mulch, but it still takes a surprising amount of water to grow vegetables.

If you want to raise animals, they will need water as well; below is a chart for what they need per day:

Animal	Gallons of Water Per Day
Beef Cattle	12
Dairy Cattle	15
Horses	12
Pigs	5
Sheep	4
Goats	3
Chickens	1 per 10 birds
Turkeys	2 per 10 birds

You can see how the numbers can really add up quickly, and I want you to be realistic, because not having enough clean water will make life hard and can pose a health risk to you, your family, and your animals.

This will be controversial, but I'd only consider a proper well for a permanent off-grid homestead as a viable option. Rainwater collection and natural springs are great for backup water sources or to supplement your well for crops or animal watering, but nothing will beat a well.

There are places where a well isn't feasible or too expensive to drill, particularly in arid areas. Again, this may not be a popular take, but if you're thinking of developing a homestead from scratch, you need to choose a place where a well is possible. Land in deserts will be cheap, which is very tempting, but you will be fighting nature at every turn.

I know some people won't heed this advice, but they do so at your own peril. Living in the outback of Australia, I saw how tough it can be when you rely on rain for drinkable water. I've also seen people haul in water by truck into a cistern, either temporarily while they save up for a well or just permanently for their water supply; it's a challenging way to live.

When living off-grid, where everything is up to you, having a redundant water source is a great way to build resiliency. Harnessing a spring or capturing rainwater can supplement your well, especially if it produces fewer gallons per minute than you'd need it to.

Rainwater capture can yield a lot of water in a short rainfall if you're set up to capture it. However, it does require you to get enough rainwater in the first place and then store it somewhere during dry spells. In my area, we get about 45 inches of rain per year, which equals 0.6 gallons per square foot of rain-capture surface.

This means that, in my area, if I had a 1,000-square-foot roof, it would receive around 27,000 gallons of rainfall per year—assuming I could capture, store, and filter that much water. You can do a quick Internet search to determine your local rainfall. This formula will help you calculate the potential for rainwater capture in your own situation.

Roof square footage × Inches of rainfall × 0.6 Gallons = Total rainfall capture

Capturing rain is the relatively easy part of rainwater harvesting; storing and filtering can be challenging. When building your system, you'll want to size it to your own situation, so calculate what you'll need from the information above and then total it all up. You'll want to add in a buffer for evaporation, spillage, times of drought, etc.

When it comes to the capture surface, you have two options: You can use existing structures or build something dedicated to rain capture. For existing structures, you could use your home, sheds, outbuildings, or even your solar panel array with the right arrangement. The biggest considerations are whether you have enough surface area to capture what you need and whether the material is safe to capture.

Avoid anything with asphalt shingles or similar products; leaching chemicals can lead to cancer if you drink from that water for long enough. The go-to material is metal roofing, TPO membranes, and PVC panels. Also avoid polycarbonate panels because they contain bisphenol A (BPA), which can lead to health issues. Even when choosing a relatively safe material, take precautions if you plan to use the water for drinking, watering your vegetable gardens, or providing water for your animals (whose meat you might eat later).

The first step is to allow the material to off-gas any volatile organic compounds (VOCs). Most of this will occur during the manufacturing process, but some will still occur for a little while after your installation. The two most significant factors for off-gassing are heat and time. As time passes, the material will continue to release chemical compounds, but realize that this is weighted very heavily to early on in its lifespan.

Broadly speaking, 90 percent of these VOCs will be off-gassed during the manufacturing process itself before you even get the material. Of the remaining amount, almost all of it will be released in the first 30 days, especially once the heat of the sun warms up that material. There will be minute trace amounts that will leach out over the remaining years, but most of that will volatilize into the air and never touch your water.

The next strategy to keep the water clean and safer is using a technique called first flush diversion. There are several ways to achieve this, but the point is to divert the first few gallons of water that wash off your roof. That is desirable because the first few gallons of water will contain a lot of dirt, dust, bird poop, and any VOCs; by using a first flush diverter, we don't let any of that dirty water enter our water storage.

After that, we will want to run our water through a coarse filter to make sure we block any animals seeking water from entering and keep out any leaves or other debris. Below that, we'll use a slightly finer filter to catch anything else that makes it through.

The next step is to have the water enter the container in a way that only disturbs the upper layers of the water that is already there. We want to avoid stirring up the bottom level of our tank because that is where sediment will have settled, and we want to keep it there as much as possible. To do this, have your water inlet either come in and take a 90-degree turn or have it fall on a flat, elevated surface to splash outward instead of downward.

When I was in the Australian Outback, I saw some clever designs to accomplish this. For more extensive storage, it was common in my area to have two chambers to your water catchment: the first as a settling tank and the second as your bulk storage tank. The settling tank slows the water down so it drops any debris or dirt.

Some of these first tanks had baffles on either the bottom or the top so that the water would spill over the first baffle, far from the sediment. Then a second baffle from the top would come down, forcing the water to slow more and drop more sediment along the way, instead of flowing straight into a pipe.

The trick was sizing: the baffle needed to be big enough to slow the water down without getting overwhelmed, but not so big that it would prevent water from flowing into the main bulk storage tank. This system of baffles would help your main tank by leaving most of that particulate matter in the first settling tank.

Speaking of that sediment, we want to use the water only after the rain has stopped and some time has passed. This lets anything in the water sink down to the bottom; no matter what you do, some stuff will get in.

Build in access so you can use a pool vacuum to clean out the bottom about twice a year. A large manhole cover is typical, but in my experience, it is not big enough. They might be big enough to crawl into, but that risks too much contamination, so we want a secure larger opening to prevent animals from crawling in and drowning

An opaque wall and cover to your storage is critical to preventing the growth of algae and other vegetation. Along with vacuuming the bottom for muck, you'll also want to treat the entire thing with bleach at least twice a year. Choose a bleach without any scents or other additives; you want only sodium hypochlorite (NaClO) or cal-cium hypochlorite ($Ca(ClO)_2$) in a 5 or 6 percent solution, which is the main ingredient that does all the work in bleach.

Use 8 ounces of bleach per 1,000 gallons, assuming that your water is clear (low turbidity). During these shock treatments, you're aiming for around four parts per million (ppm). In between, you should test to ensure that you keep this level between 0.2 and 1.0 ppm. If you've previously shocked the system with the four ppm treatment, you monitor the water regularly for its levels and can add 2 ounces of bleach per 1,000 gallons to keep these levels up.

When shocking your system, distribute the bleach evenly and use something clean to stir and agitate the water (after vacuuming the sludge) to incorporate the chemical thoroughly. Wait an hour, and then go into your home and run a tap long enough that water from the treated cistern has been drawn in. You should be able to smell a faint level of chlorine. If you don't smell anything and you're sure that the tap has run long enough to get the treated water in again, you'll want to add more bleach to the cistern. Wait another hour and repeat until you can detect a faint chlorine smell.

At that point, go through your entire house and run every sink, shower, washer, dishwasher, and hose bib until you smell the chlorine. Then, turn everything off and let it sit for a few hours. This will not only treat your water source, but will disinfect your plumbing system, too.

Chlorine in this small dilution will keep your water safe and is considered below a threshold that can do harm. Because of its chemical nature, chlorine will dissipate through the water and evaporate over time. The bulk of the chlorine in a shock treatment will dissipate in about a week, but I don't hesitate to drink the water after a few hours or a day at most.

The last part of your system will be a water pump to deliver the water from your cistern to your point of use. Flowing from the tank, you'll want to filter everything. Your first filter should filter the water for any sediments pulled in; look for a filter sized at 10 microns.

Your second filter will reduce sediment further to 5 microns. We do this to remove particles in stages to prevent premature clogging of filters. This is also an essential step in reducing any turbidity or sediment that makes the water opaque, which will be vital in the final stage of our filtering.

Next, we have a charcoal filter at around 5 microns, which helps improve the taste of the water, absorbing and further filtering out anything that we don't want in the water. It's important to note that when you first get a charcoal filter, it will likely produce dark water as loose charcoal is being rinsed out. Flush the filter until the water has turned clear, which is important again to reduce turbidity.

The last stage of our filter is a UV light filter. This is a metal chamber with a UV light in the middle of it. These filters are very effective and have been used for decades, with the added benefit of not utilizing any chemicals. They are pretty low maintenance. The water flows into the chamber and is exposed to the light, which destroys any bacteria, viruses, or microorganisms.

You'll need to choose a UV filter sized to handle the amount of water you'll need at any given time, rated in gallons per minute. This is because you want the water to be exposed to the UV light long enough to fully destroy anything undesirable. This is also why it's important that your water be totally clear at this point. Any turbidity can reduce the effectiveness of a UV filter.

Make sure to follow the manufacturer's directions on maintenance, testing, and filter changes in order to keep your system functioning properly so that it can protect you. I have seen what giardia can do to a person; you want to take your water very seriously.

Another water source you can develop are natural springs. These can be fickle things, sometimes being seasonal, sometimes just stopping permanently for reasons beyond our knowledge. That poses a real challenge when you rely on that spring as your sole water source, which its why many banks will not underwrite mortgages for homes on a spring.

If you're fortunate to find a natural spring, then it's worth developing. The hardest part is finding the original spring head without disturbing the soil around it too much. Once you do locate it, you'll want to install a spring dam wall, which is a thick piece of plastic that curves around the spring head to collect the water and keep it protected.

Your spring dam wall will typically be about 12–18 inches tall and about 6–8 feet across. You'll clear as much debris and dirt around the spring head as you can, then install the dam wall so that the ends curl into the hill side above the spring head itself.

This allows the water to start pooling behind the dam wall, where you'll install a pipe through the center point of the dam wall, just above the bottom, so clear water can flow into it and out the front pipe. You want this pipe to extend away from the dam wall and out to your water catchment so that you preserve the water quality and not give the water a chance to wash out the dam wall you built.

Some people use hydraulic cement to seal the floor behind the dam to prevent it from washing out and also to make it easier to clean it out better. You'll also want to install an overflow valve that can divert any extra water that the spring might produce so it doesn't wash out your spring head dam.

Finally, you'll install a PVC pipe that is positioned for you to pour bleach into the space behind the dam to sanitize it from time to time. You'll then fill everything behind the dam with washed gravel, ensuring that you don't damage the pipes and that you keep them at the right heights. Finally, cover the entire area with a 30-mil pond liner to prevent any dirt, bugs, or debris from falling into the gravel.

While natural springs and water capture are great supplements, your best option is to drill a well, which is one of the most important investments you'll make on your homestead. Drilling a well is neither inexpensive nor for the faint of heart. There is a lot you can do to control your own destiny; with enough hard work, you can make almost anything happen. But drilling a well is one of those rare cases when you simply need lady luck on your side.

If water is not flowing, you can always drill deeper or drill another hole, but most of us don't have unlimited funds to keep searching for water. If you can count on one thing when it comes to well drilling, it's that it will be expensive. That said, compromising on water in any way is going to set you up for a very difficult life, if not failure. Water is vitally important.

If you remember back to my land development order of operations, first develop access, then drill the well. Ideally, I would do the reverse, but you need to be able to access the land before you can drill it.

Drilling a well starts with making sure that the well rig can get where it needs to go to set up without any hassle. These are large trucks that are loaded with heavy scaffolding and drill pipes so you can't expect them to go off-road. Also, the harder it is for them to access, the more money they'll charge you.

With that in mind, I'd rather spend dollars on making a nice driveway and keeping drilling costs lower than skimping on my driveway and paying extra because the well-drilling company tacks on additional costs for the hassle. In the end, I'd like to keep my dollars on my property in the form of a driveway instead of rolling away in a well driller's pocket.

Videos and online discussions about hand-drilling a well yourself with simple PVC setups make it temping to tackle this on your own. However, this is one of those cases where it's too good to be true. The biggest problem is you won't be able to get a certificate of occupancy from your local municipality, meaning that if you live in the house and the authorities find out, they can fine you and condemn the house.

Also, these hand-drilled wells don't penetrate the bedrock, which means you'll only have access to the water floating above the bedrock, which is more suspectable to contamination. So forgo the temptation, health risk, and heartache that these makeshift wells will likely bring and develop a high-quality water source for your off-grid homestead.

When it comes to choosing a site to drill, there are a lot of competing opinions and thoughts on the matter. Your options include consulting local and county resources like hydrology maps, bringing in a geologist, or even giving water dowsing a try despite its dubious claims. In the end, I'd suggest choosing a well driller that has a good local reputation, and then trusting in their experience.

You want to drill your well in a place relatively close to your homesite, that's easy to access, and isn't going to be subject to flooding or similar events. It's important to note that your well should be far enough from your homesite to not interfere with construction, but close enough that you don't have to trench and lay pipe too far. Each foot of trenching is expensive or hard work.

The well drilling process is fairly complex and requires expensive equipment, but we'll review the basic steps. First, the company will drill through the top layer of soil, sand, clay, and rocks until they hit bedrock. At that point they'll install a metal casing from the surface down to the bedrock. They'll then drive the bottom of that metal casing down into the bedrock to thoroughly embed it into that layer.

This casing allows you to maintain that bore hole through this layer of soil, sand, or clay so that it doesn't collapse or get contaminated. You're essentially building a protected tube from the surface to the bedrock.

They'll typically drill a hole that's a little bit bigger than the metal casing, then fill the space around the outside of the casing with some sort of cement or bentonite. This further protects the casing and makes sure that nothing from the surface has a fast track to the water table in our bedrock.

After that, they'll continue to drill down into bedrock through that casing. The goal is to drill far enough that they intersect enough factures in the bedrock to create adequate water flow. Those fractures fill with water from the underground aquifer, and if you cross one that is big enough or a number of smaller ones, that will give you enough gallons per minute.

You're hoping for at least five gallons per minute in this process. You can make due with less, but then you get into larger holding tanks, extra costs, and having to watch how much water you're using. On the other hand, you can also end up with a lot more water, which would be a blessing.

From there, the drilling company will drop in a well pump that is wired and connected to the surface with a pipe. One thing to keep in mind is that if you're on solar, your inverter must be able to support 220/240 volts because most well pumps run on that.

You also need to understand the surge power requirements for your pump, because the pump will draw up to three times its running power consumption in the first one to three seconds. Most good-quality inverters can handle these types of spikes, particularly if they're able to provide 220/240 volts. You can check the specification sheet for surge power ratings to be sure before you make any decisions.

Typically, once the water comes out of the ground you'll fill a pressure tank to give your home the water pressure it needs. Along with that, you might add water softeners, water filters, etc.

Whatever your method for getting water to your home—whether springs, wells, water catchment, or all of the above—testing is important. You need to know the quality and contents of your water. Some tests will be required by your local government, and others are just a good practice. Water testing is complicated and there are some less-than-honest private labs out there. In general, though, most states have their own labs or certify others to ensure you can trust the results.

You can do certain tests on your own, while other tests have to be done in a lab. Sometimes the water must be collected by a trained person in order to minimize errors or to stabilize the water content for it to be accurate by the time it gets to a lab. The best way to do this is connect with your local health department; they'll be able to point you to reliable resources, let you know what things to look out for, and advise you on the specific tests that are important for your area. After all, you don't just want a reliable source of water; you need a reliable source that is also safe.

STEP 40   HANDLE GRAY WATER

—

Once you get your clean water out of the ground and use it, you'll need to figure out where it goes next. The answer depends on which of two categories the water falls into: gray water or black water. Gray water is wastewater that isn't clean enough to drink but isn't dirty enough to be considered sewage (which is black water).

If you have a septic, you essentially have a great system for dealing with gray water. The septic tank handles the solids and then the leach field is just a run of tubing the spreads out the water to be absorbed into the earth in a controlled manner.

You might decide to use a composting toilet, or you may have a septic, but you might also decide to divert some of your gray water for landscaping or maybe your garden. Remember, this water is barely dirty, so in many cases it's fine to use in these areas.

Diverting gray water is an ideal solution for managing wastewater in off-grid homes. In my off-grid tiny home, I use a modified French drain system for gray water since I don't generate much wastewater. However, if you are building your home and you have several people or a whole family, you'll need to consider the volume of wastewater you're producing, because it won't be a trivial amount. If you remember back to earlier in the book (step 17), you estimated how much water you need for drinking, cooking, cleaning, etc. You can leverage your septic for the bulk of your wastewater disposal and divert a portion of it.

Now that you know how much water you plan to put into your gray-water system, you'll need to figure out where that water is going. It should be at least 100 feet away from your wellhead and your septic field. It should also be at least 20 feet from your home's foundation unless it's a really small amount.

By keeping this water away from your well, you won't run any chance of contaminating your drinking water. You'll also want to avoid your septic and leach field because that soil is already working to drain wastewater, so adding more might overburden it. Finally, keeping gray water away from your foundation will help protect your home's integrity and reduce moisture issues in your basement.

A key step is to determine how you are going to remove gray water from your house; you can't just use the normal drain lines since they also contain black water. Some people run an extra drain line from their sinks or washer outside into the garden.

Much like your water catchment system, you'll want to do two things before you let that water enter your dispersal water lines: slow it down and remove any particulates. Slowing it down involves installing an in-ground holding tank that's sized to handle the maximum amount of water you might divert to it at any given time.

For me, showers were the biggest producer of gray water. My shower head used about 1.5 gallons per minute and I take about a 10-minute shower. With that in mind, I sized my holding tank to 20 gallons, which gave me some buffer. An important thing to remember is that you'll need to bring in pipes above the water level, so size your tank accordingly.

The water enters my holding tank near the top through a 1/2-inch-diameter PVC pipe (which is big enough to handle 14 gallons per minute). After this pipe enters the tank, I made a 90-degree turn and pointed it downward. Then, on the other side of the tank, I brought in my exit line, which was positioned to be about two inches lower than my inlet, but this time, I had a 90-degree turn pointing upward.

A key detail here is that the exit pipe was one inch in diameter, which allows for more than double the flow rate as the inlet. In this way, you never have to worry about water flowing in faster than it is flowing out, potentially backing up into your home.

The tank that I chose also had slots in it to insert a media filter pad before the water exited the tank. This helps achieve the second requirement of removing particulates. The aim here isn't necessarily to clean the water but to prevent particles from gumming up your water lines and outlet ports in your French drain pits.

The biggest culprit here typically is food bits if you divert water from your kitchen sink. Similarly, your clothes washer and bathroom sinks contain small particles of skin, dirt, and debris from washing your hands or your clothes, so we want to capture that in our filters.

Once the water exits from your surge tank, you'll then want to pipe it below your frost line to the location where you'll let it soak into the landscape. When deciding where to deposit this gray water, you can bring it to your gardening or landscaping, but you want to make sure that the soil there drains well.

You can check this by performing a simple perc test in the spot where you're thinking of dispersing your gray water. If the soil drains well, then it is a good candidate for handling gray water. Estimate the amount of gray water you'll produce daily and design your system to handle that amount plus an extra 25 percent.

If you have a lot of water, I suggest splitting your lines so that they deposit the water in multiple locations. For example, if you notice that your soil can handle one gallon per minute, but you're producing a peak of seven gallons per minute, you'll need seven dispersal sites, plus two more for good measure.

Space these sites well away from each other so that the soils can absorb the water easily. As you install these branches, make sure that your drain lines are downhill from your point of use, and dig trenches deeper as needed to achieve a slope for proper drainage.

You want to make sure that your water lines have a quarter inch of drop per foot of run, and ensure that there aren't any dips in the line. Otherwise, the water will sit there, and any particulates will settle. Having water sitting in a pipe can also lead it to freeze in winter months. The best way I've found to achieve this is to trench about an inch deeper than you need, then pour in a thin layer of sand. This will let you make micro adjustments easily along the run of the pipe.

Once the pipes bring the water to the dispersal location, dig a hole twice the size of the volume of water you're expecting it to handle. I tend to size these gray-water pits along the lines of a five-gallon bucket size and shape (keeping in mind how much water my soil can handle), because it's one of the most useful tools on my homestead.

I dig a hole, putting the dirt I dig out into a bucket or two. I also dig down about 3 inches around the opening of that hole in all directions to later insert my cover. I'm shooting to get below the frost line in my area by about 12 inches, so that the water has time to soak instead of freezing. Then I take a five-gallon bucket of washed gravel and fill that hole to about one inch below frost line.

From there, I bring in my pipe at the frost line, where it will deposit the water in the middle of the hole. At this point I use some landscape fabric to cover the area and prevent any surface dirt from entering my pit, add a thin layer of gravel, and then install a landscaping water valve box just above the surface.

Around the box I lay landscape fabric to keep weeds at bay and additional gravel to keep the box down and dirt away from the opening. The last step, for colder climates, is to stuff the empty water valve box full of fiberglass insulation bats. This helps to keep out the cold, since you've essentially created a depression in the land which cold will settle into.

Create as many of these as you need to handle your system's water, adding a few extras in case one gets clogged or you have guests over and you're producing more wastewater than normal. At this point, you have a fully functioning gray-water system.

Keep in mind that dispersing your gray water requires you to carefully consider your soaps, detergents, and cleaning supplies. Your water carries these substances into the soil, so it's important to know what they contain. While some soaps naturally degrade in the soil pretty quickly, with minimal harm to the ecosystem, others can persist for a long time and possibly do a lot of damage. When it comes to soaps, the two brands that I've relied on over the years are Dr. Bronner's Castile soap and Dawn dish soap. Dawn has its detractors, but I've found that it strikes a good balance of working very well and being easy on the soils.

Along with personal care products, you should also consider your household cleaners. I've largely switched from commercial cleaning products to simple white vinegar and baking soda. There are times that I reach for stronger chemicals for specific situations, but that's more the exception than the rule.

Unfortunately, there is no perfect solution—despite what people online will tell you—but being mindful of your products can help protect your land and your local environment. You should be particularly careful not to divert gray water that contains chemicals to your vegetable gardens. Ornamental landscaping is less of a concern, as long as you don't dump a lot of salt down your drain. While most people never think of where their wastewater goes, part of living off grid is understanding both your inputs and outputs.

STEP 41   HANDLE BLACK WATER

—

When water gets too dirty to reuse, it becomes black water, also known as sewage. This water comes primarily from your toilets, which means that you need to be particularly careful in how you handle it. The ability to handle black water safely, along with access to clean water, is a major struggle for some of the world's poorest countries, leading to dangerous sanitation and health problems. In short, black water is a serious issue.

Next to your well, a septic system is the most important part of your infrastructure you'll need to set up—and that does come at a cost. You should have had a perc test before you closed on your land, and with it, a septic permit in your own name. Completing that test and getting a permit in place removes a lot of risk from this process, which is something that can't be said about drilling a well.

Each municipality has its own rules and details, but in most places, your septic field needs to be at least 100 feet from your well and 150-plus feet from any natural body of water. It's worth noting that, in most areas, it's illegal to have an outhouse or pit toilet.

Composting toilets are a bit of a gray area, and your mileage may vary depending on which toilet you use and where you are. Where I lived, the city allowed composting toilets as long as the contents were bagged and put into the trash pickup bin. At that point, it was equivalent to tossing a baby diaper in their eyes. The alternative was to tie into the city sewage line, which would have cost me an extra $50,000.

For most of us, this all adds up to needing a septic field. In the county where I built my off-grid homestead, regulations required me to have the septic installed by a licensed contractor. I could have gone through the certification course and done the work myself, but in the end, the cost savings would have been offset by the licensing fees.

Fortunately, the county maintained a list of all the companies that had an active license with them, and with this handy information, I started calling. Along the way, I cross-referenced the companies with online searches to make sure they were well reviewed, since I couldn't find any recommendations from the people I knew locally. The process went smoothly from there. I scheduled the work, met the company on the land, and as I observed them in action, it became clear I had made a good choice.

First, the workers set up a laser level to get an understanding of the grade of the land. Your system needs a quarter-inch slope per foot of run from the house to the tank, to the entrance of your leach field. The field itself will likely require several legs, which will follow the contour of the land. It is very important that you slope things or keep them level in the right spots, without having any dips where sewage will pool.

Once the workers mapped out the slopes, they started by digging the main pit, which indicated the topmost point of the system. They then lowered in the septic tank, which was a cement casket, and leveled it in the hole. At that point, they measured the outlet port, which became their "zero point." From that zero, they could measure each slope to make sure the angles were correct. They then dug downhill until they got to the split in the line from which each of the leach field legs branched.

The team that installed my septic was clearly well practiced at this balancing act. As one worker dug out the trench, another used the laser level rod to carefully measure the slope of different parts of the landscape. The worker using the backhoe was also quite skilled, so very little needed to be dug by hand.

As they moved along, continuously checking the slope, the worker with the laser level rod used a shovel to level out minor imperfections and prevent dips. Over all, I don't think he had to make more than 20 scoops of dirt the entire way.

They then began to lay in all the PVC pipe according to country requirements, and started attaching the leach field tubing. This tubing was heavy-duty drain line with a layer of foam packing peanuts held in place with a filter fabric sock. They resembled funny-looking hot dogs that connect to create the leach field.

Once all the pieces were assembled, the workers started back at the top of the septic and checked the level off the tops of the pipes to make minor adjustments. Then they had the county inspector come out and check their work.

He started by checking the settings of the laser level, then measured the slope himself at each point in the system. He made sure that all the connections were glued properly and that the right-size PVC was used, and he measured the leach field length. Finally, he used a GPS unit to measure the coordinates of the tank for the county records.

The key thing about this inspection was that, because it was "open trench," every foot of the system had to be uncovered and visible. So make sure you get your inspection *before* you fill in anything, or the inspector will make you dig it all out and start again.

In addition to your leach field, most municipalities require you to have enough space for a backup field. This way, if the first one goes bad, you'll have a place to put in a new field and the homestead won't be at risk of not having a septic. Typically, this is determined when you apply for the permit; you'll mark it on a map of the property as part of the application.

With a backup leach field, and proper grading of the first leach field, you should be in good shape to handle your black water safely and easily.

STEP 42   WASH LAUNDRY OFF-GRID

—

When I first went off the grid, I hadn't realized how much laundry was going to be part of my life. I had backpacked for years, sometimes needing to wash clothes in a sink, and had spent hours at a laundromat, but living off-grid was different.

The average homesteader in the 1900s spent a full day and sometimes two days per week washing, drying, and folding laundry. And that was in an era when one person worked at home full-time, whereas today both adults in many families work outside the home. With the miracle of laundry washers and dryers, the average household now spends 85 minutes per week on laundry. This all adds up to you having a lot more time to enjoy your homestead instead of spending many hours just keeping up with your chores.

When I first went off-grid I was going into an office every day, and that meant that my clothes had to be clean, pressed, and neat-looking. That was no small feat when I was trying to dry on a clothesline, not to mention the time when I was in dress slacks and loafers, wielding a chain saw to clear my driveway of a fallen tree at 7:00 a.m. just to get to work on time.

Laundry on a homestead is a never-ending task, especially when you're rolling up your sleeves to get real work done. Dirt, grime, stains, and all sorts of things from your animals add up to homesteaders having dirty clothes.

The real challenge when it comes to laundry is drying the clothes. Many people have idyllic visions of warm spring days with a slight breeze, hanging your bed sheets on a line in the morning and falling asleep to that fresh smell at night.

What most people don't think about is when you're rushed and don't get to the wash until late in the day, but it's so humid that the laundry sits on the line for hours and never gets any drier. In my climate, it is very humid in the summers and relatively rainy in the winter. Both types of weather are a challenge when you're hang-drying your laundry.

Clothes dryers can also be a challenge when you're living off-grid because they are very power-hungry and can run for hours if you have a family or a lot of clothes. The average modern clothes dryer will use about 3,000 watts when running; that can be a heavy tax on your power system.

One alternative is to use a gas-powered dryer, which takes those 3,000 watts down to about 200 watts (plus the gas you'll use). This is another case in which I feel that using propane or natural gas is really your only practical alternative. A gas-heated dryer typically adds about 20 percent more in cost, plus the installation cost of the gas line. That said, it will likely save you thousands of dollars in not needing additional solar panels and batteries. So going with a gas-powered dryer is my recommendation when it comes to doing laundry off the grid.

That doesn't mean you can't still use a clothesline; it's worth saving the money if the weather is pleasant. Many homesteaders still use a wash basin, scrub board, clothes wringer, and clothesline to do their laundry. You might start there while you're saving up for a more modern solution of laundry appliances and solar. You'll also likely develop a deep appreciation for those modern conveniences.

Washing can be as simple as putting a drain plug in your kitchen sink and filling it with warm water and a little soap. When backpacking, I did this a lot; I'd use a bit of Fels-Naptha soap, spot-treat the clothes, put them into the water to soak for about 15 minutes and agitate the water with my fingers for a minute every once in a while. I'd then drain the water, wring out the item, rinse it, and soak it twice. This did a pretty good job, but after a few washes, the clothes needed a deeper clean.

For larger loads and a better clean I tend to use large wash bins, with the same process, but I add in a washing plunger. These are amazing to help push water through the weave of your fabrics and remove the dirt embedded in them. The plungers look similar to a toilet plunger but operate in an entirely different way.

Modern ones help drive the water through your fabrics with a cone that agitates the water with more force than you could exert with your bare hands. They also oxygenate the water in the process, which can help with smells and stains. You'll want to think about ergonomics in this process because between lugging a lot of water and plunging your wash plunger, it's going to take a fair bit of elbow grease.

The next most important tool in doing your laundry is going to be a clothes wringer roller, which will squeeze the water out of your clothes. This will help between rinses and dramatically improve your ability to get your clothes cleaner than with just hand-wringing. It also cuts down on dryer times.

I've also heard good reviews of manually powered or low-power laundry spinners, which can remove more water. The downside is they take up a fair bit of space, they cost more (if you can find them), and use power. Roller-style wringers will do a good job on most laundry but struggle to wring the water out around items like buttons, zippers, and clothes that aren't of even thickness.

For those who live in a very cold or rainy climate, you'll want to take into consideration how you're going to dry your clothes when the weather is bad. Clothes don't dry very well when they freeze solid or it's so humid that the clothes can't quite get dry.

This might even mean devoting space in your home to drying clothes. I knew a homemaker who cleaned up an old outbuilding and ran clotheslines through it as a dedicated space to dry her family's clothes in the winter. In addition to the clotheslines, she had a woodstove in the corner that was oversized for the space. She'd get the fire started, hang all the clothes, then shut the door to the building, and it would heat up to a very warm temperature. This meant she could dry her clothes "on the line" no matter the weather.

Having a separate space to wash and dry your clothes when using more traditional methods can be a big advantage. When I lived in Europe without a dryer, like most Europeans do, I had a drying rack constantly set up in my living room. That was fine for a while, but when it became a permanent fixture just to keep up with the laundry, I soon wished I had a dryer or at least a separate space to dry my clothes.

Another option is to utilize a laundromat for a certain period or even long-term. When I sat down to think about what type of life I wanted to live, imagining my ideal day, one of the things I wrote down was "not have to do laundry." When I added up the cost of a washer and dryer, the solar panels and batteries to run them, plus the cost of a plumber to run my lines, it totaled in the thousands.

Comparing that to the cost of a wash, dry, fold at my local laundromat, I could pay to have my laundry done weekly for 7.8 years before it eclipsed the up-front costs. Add to that, I didn't have to fold another shirt and I got all my time back; it was an easy decision for me.

STEP 43   COOK OFF-GRID

—

The heart of a homestead kitchen is the pantry: the place to keep your deep stores of preserved food. While this is a challenge to find in many homes today, a large pantry is essential.

When you live off the grid, you will likely want to grow a good portion of your own food—and you will always need to store food in case you get snowed in or live far away from grocery stores. In my area, the government doesn't plow the roads when it snows. That typically means that I'm cloistered away on my homestead for at least a few days. Living in the country also means you can't just swing by the store; I've lived in places where the closest grocery store was almost an hour away. This meant that I needed to keep a pantry to have what I needed on hand.

If you have the opportunity to design and build your own home, do your best to choose a design that allows for a large pantry. You can supplement that with storage in a root cellar, basement, or other location, but nothing beats having all your ingredients and food right off the main kitchen. When it comes to pantry storage, it's hard not to agree with the truism that more is better. Consider how many cans and dried goods you want to store, and whether the space will also house crockery and countertop appliances.

My advice is to use shelves that aren't too deep. I find that once my cans go more than 3 deep, things start to get lost. Ball jars are 3 inches wide for 16 ounces and 3.4 inches wide for 32 ounces, so a 12-inch shelf works well to fit three rows deep. Using shallow shelves requires more wall space, but it helps you make sure you are rotating your cans more effectively and safely.

The ideal pantry will also include deeper shelves, typically around 24 inches, for larger pots and appliances. I tend to store these types of items in base cabinets topped with a counter for additional work-space. On the other hand, I place cans at eye level so I can easily see back into the shelves, which means installing shelves higher on the walls, with some dry goods stored below them in larger bins.

Bringing power into your pantry is also a nice bonus. I found that additional countertop space in my pantry let me keep a few things like a toaster or stand mixer out and ready to use, but not cluttering up my kitchen's countertops. Power in your pantry is also helpful for things like bread makers, slow cooker, or plug-in water canners that need to run for a while. It's great to have a place to set them up and let them do their thing.

One drawback of a large homestead pantry is pests. Since you have a lot of food in a single place, you run the risk of attracting unwelcome visitors. You'll typically face four main types: moths, weevils, beetles, and roaches; mice are a concern, too. Some of these pests will come in the food you buy from the store, others might arrive from the garden on produce you grow yourself, and still others will find their way into your pantry on their own. Over the years, I've found that a few steps can really help keep these critters at bay.

First, start with a well-maintained space. Before you fill the pantry, clean all your surfaces, check for eggs, and caulk every little nook and cranny to physically exclude pests from the space. This also goes for your house generally; the harder it is to get into your home, the less likely that pests will get to your pantry.

Next, I suggest painting the walls a bright, light color and installing good lighting so you can see pests easily should they come into the pantry. A fresh coat of paint will help you spot a cluster of eggs, a dead ant, or other evidence that bugs are around.

After that, make sure that you're properly managing your pantry, storing only what you'll use in a reasonable amount of time. Practice FIFO: first in, first out, putting the oldest items at the front of your shelves so you use them first. Between seasons, wash your entire pantry and check for evidence of bugs. Don't forget to inspect the bottoms of your shelves, too.

The biggest vector for weevils, moths, and beetles is actually direct from the store. These pests are also among the more destructive things that can afflict an otherwise well-managed pantry. The best protocol I've found is to purchase bulk items that have been vacuum sealed with oxygen absorbers in the bags. Once I bring them home, I put them directly into my chest freezer for at least three days.

Freezing bulk goods like flour, rice, sugar, salt, and beans will kill these bugs at all stages of their lives: the eggs, larvae, and adults. To do this, I keep a small chest freezer dedicated to this task in my garage. This means that when I get home from the store, these items don't ever come into my home before being frozen. They move directly from my car's trunk into the freezer; then, after three days, I move them into my home pantry.

If you're purchasing items that are sealed properly, you won't have to worry about any additional moisture being introduced, and freezing these types of items doesn't harm them in any way, but it will harm those bugs. If the items are loose, I use a vacuum sealer to pack them in sizes that are convenient to how much I use that ingredient. I keep a supply of individually wrapped oxygen absorbers to add as an additional layer of security against pests. Keeping your items in smaller quantities will also help limit the possible spread to just those containers that are opened. These steps will keep bugs out of your pantry, for the most part.

Unfortunately, mice are a different story because they find their way into almost anywhere, and they can chew through food packaging. For this reason, if I store anything on the floor, it's in a metal container with a tight lid. Even if you build a new house on your land, you may still have a mouse issue because they live nearby. Your construction has displaced their home, so they go to the next best thing, which typically means your house. For this reason, I like to keep the woods cut back from my house a little bit and mowed grass or mulch between. It's also worth hiring a pest control service, regardless of whether your home is old or new.

Beyond right-sizing your pantry and keeping out the pests, two other important things to consider are making sure your power for your fridge and freezer stays on, and being able to handle canning season for your food preservation. These two issues may seem simple but they turned out to be bigger tasks than I realized when I first went off the grid, which led me to rethink how to best approach them.

While living off the grid, you may experience times when your batteries bottom out, you need to take your house offline while you do maintenance, or, if you use the grid at times, a storm knocks out power. Whatever the cause, you need to keep the household running.

From the homesteader's perspective, the biggest challenge when the power goes out is preserving food. You may have just bought an entire cow for meat for the year, spending a lot of money, and a power outage can jeopardize your bounty. Similarly, you might have worked hard to raise a flock of broilers, and they're sitting in your freezer to eat throughout the season.

This is when having a whole-house generator powered by natural gas or propane comes in handy. These devices can sense when the power goes out and turn on automatically to keep your fridge cold and your house functioning. A freezer can keep your food safe for about a day of no power if you don't open it. I also take the step to have a WIFI-enabled temperature monitor that will send me an email if my freezer gets too warm. This is great peace of mind not just for power outages, but also for a tripped breaker, or if your freezer just happens to go bad.

When going on vacation, I take the extra step of adding an extra thermometer with a long probe. The probe goes inside, the screen goes on the outside, and I put a security camera facing the screen, so I always have two ways to check the temp. I've set up my WIFI to be able to automatically reset if it can't get a signal for more than 15 minutes. If the temp were to ever get too warm, I can phone a friend with a key to see what's going on when I'm away from the farmstead.

The second thing I learned over the years is to have plenty of space to preserve foods during the harvest. Canning typically happens when you have large flushes of produce. While I stagger my plantings in the garden across a few weeks so that I eat my produce as I harvest it, the exception is those items I want to can. I plant them all at the same time so that I can harvest them all at the same time.

Your preferences will dictate what you can, but for me, tomatoes are an obvious choice for sauces and salsas. I do cucumbers for pickles, too. When I plant a lot of tomatoes, they come in quickly and I plan to can in large batches.

Canning can take up a lot of space between heating all your water, laying out your jars, processing the produce, letting your processed cans cool down, etc. Sometimes you can do the work right in your kitchen. Other times, it makes sense to set up a temporary canning kitchen to expand your work area. Canning kitchens were traditionally built outside the home in a semi-enclosed space to help maintain sanitation while allowing heat to dissipate. In older times, keeping a kitchen cool was a big deal; today, with proper venting and the miracle of AC, it's much less of an issue.

Still, if you're building your own home, consider venting your range, directing the vent outside and using a hood that is rated to draw a lot of CFMs (cubic feet per minute). Between keeping the heat down and directing the bacon grease smells out, it's worth the extra materials to run the venting tube to the outside.

Some homesteaders also build an outdoor kitchen as part of their patio or in a clean part of an outbuilding, bringing in a bunch of folding tables for a big canning session. Having many propane burners allows you to boil lots of water at the same time.

In the past, I've organized canning parties using a local church's kitchen; they had a large commercial-grade kitchen for community meals. We rented out the kitchen for a day, all the participants chipped in a little bit, and we required everyone to bring 50 pounds of tomatoes they'd grown along with a new pack of jars and lids.

Everything went into the same pot, and everyone helped skin, chop, and process the tomatoes. In the end, we split the jars up evenly and we all worked to clean up with the help of a commercial dishwasher that only takes a few minutes per cycle. It was a lot of fun to chat with others while we got the work done, aided by a kitchen purposefully built for efficiency.

You'll need to figure out how much space is required to store all this food, and to design your kitchen pantry and other areas around that. You will also want space to store the preserving equipment, including pressure canners, empty jars, dehydrators, giant soup pots, slow cooker, and folding tables.

My advice is that you should make your best estimate about the necessary space and then double that number. You won't regret having more space, and it will help you keep things organized better, work in the space more easily, and add a buffer if you decide to grow more in the future.

With your food properly stored, you can turn to cooking itself. Preparing meals while living off the grid really isn't much different than cooking on the grid, though many will choose a gas stove over a standard electric. If you really want to be self-sufficient, you might consider a wood-fire range, which will require some practice.

There are also a few items that are specific to an off-grid homestead kitchen that you might not find in a traditional kitchen. Along with preserving food for storage, you may want to handle larger animals from a hunt or slaughter, cook without electricity, or take up other activities.

One of my favorite ways to cook off the grid is with a solar oven. I got turned on to these when I wanted to use a slow cooker while being on 100 percent solar and realized it was too great a power drain. A solar oven concentrates the sun's heat into a hot box where it can cook foods.

Many DIY plans are available, but after trying a few of them and testing many commercially made ones, I realized that the store-bought models were far more effective and easier to use. I won't go into specific brands, but you can find reviews online that will help you decide which is right for you.

I found that solar ovens that use a vacuum tube as the cooking chamber were the fastest, but they had the downside of having less internal volume. For just me, these are great; for a family or if you want to cook a whole chicken, the sealed-box style is more practical.

What's good about solar ovens is that they operate much like a slow cooker, so any recipe for that appliance will also work wonderfully in these. You just need to make sure you're cooking on a sunny day, mostly free of clouds. Solar ovens can't be use all the time—only when you have lots of good solar exposure.

The night before I plan to do my solar cooking, I prep my food in an enameled pot with a lid, keeping it in the fridge for the next morning, when I set up my solar oven. Once I drop the pot in the chamber, sealing it up tightly, I can get on with my chores and work around the farm. By the end of the day or sometimes by lunch, everything is perfectly cooked without any electricity.

Another key tool for off-gridders is a chest freezer. These are affordable, hold a lot of food, and are perfectly suited for living on solar. Chest freezers are very well insulated, and because you open the lid from the top, they hold that cold very well, even when you open them up.

Freezers and fridges with doors on the front are very convenient, but each time you open the door, the cold air spills down and out of the compartment. They aren't any less energy-efficient, but the compressor does run more often, which in turn uses more power.

Fridges and freezers are generally easy to power from solar panels; around four solar panels (280 watts each) and two 100-amp-hour batteries will easily power a single fridge or freezer under normal use. That also accounts for some cloudy days and assumes that your location gets about three solar hours per day year-round.

Another hallmark of a homestead kitchen is cast iron. I'll go against the crowd on this one and say that I've used it, it has its purpose, but I'm not a huge fan of cast iron in a modern kitchen. Now, I will concede that cast iron has a lot of benefits. For instance, it is very good at holding heat, preventing a dip in pan temperature when you add cold foods like meat; it can be great to sear steaks! The other real benefit of old-fashioned cast iron is that the chemical coatings in Teflon and similar pans have been linked to health issues, including cancer.

Some will say that cast iron is naturally nonstick when you season and cook with it properly. While I could achieve these results after getting the hang of things, I ended up going back to my normal pans. This is what I mean by figuring out what is right for you; suggesting that you don't like cast iron might seem like sacrilege in some off-grid homesteading circles. For many, it's great. For me, it wasn't. All that's important is whether something works for you.

The last thing I will say about cast iron is that while I still use it in specific cases, I largely keep it around for aesthetic purposes rather than cooking, outside of my enameled cast-iron Dutch oven. My Dutch oven is a key piece of equipment for roasts, soups, and breads. The main difference is that it's enameled, which I find helps a lot.

A bread machine is another great small appliance to consider. While I wish I could say that I make a fresh loaf of sourdough each week from scratch, kneading by hand to produce a wonderful rustic loaf, I can't. Life is just too busy for me to do all the proofing, kneading, and keeping my cultures alive.

Comparatively, a bread machine is a great compromise when you're a busy professional who wants to stay close to the roots of your food but doesn't have a lot of extra time. These appliances can produce a good loaf reliably from wheat you grow or just grind yourself. They also have built-in timers, so your loaf can be ready when you get home from work or come in from the fields.

To go along with that, a grain mill can be a great addition if you want to mill your own flour from wheat berries. This lets you have more control over the types of wheat you use, and you can mill the berries right before using them to help preserve the nutrients.

With the electric versions, you can mill flour in a matter of seconds without any manual cranking. When stored properly in a pantry, the wheat berries can last for 30 years, which means that you're able to buy in bulk. Just remember to freeze them before bringing them inside, since these are prime candidates for weevils.

Another gadget that can be helpful is a mandoline slicer. When you have to prep a lot of vegetables for a soup or for canning, a mandoline can make quick work of a pile of onions or almost any other veggie. Mandoline slicers are super affordable and can save you a lot of time.

A more specific cutter is an apple peeler, corer, and slicer tool. These make quick work of peeling apples for pie filling, dehydrating apple slices, or prepping for apple sauce or apple leathers. I'm generally not a big proponent of single-function items, but this one makes the cut if you like apples as much as I do.

A small compost bin is another great addition to a homestead kitchen, because your greatest source of raw materials will come from your leftover food scraps. You can put it right on the counter while you're peeling vegetables. Keeping it small also means that you must empty it more often, which is actually good because it keeps smells to a minimum.

A good set of knives is important in a kitchen where you need to get work done. When it comes to knives, I'm a minimalist with only three varieties: a chef's knife, a paring knife, and a long scalloped bread knife. You commonly see sets sold filled with a bunch of low-quality knives; instead purchase each knife individually.

Plan to spend several hundred dollars for your three knives; this is one of those buy-once-cry-once items. Many good brands are available, so don't get sucked into the hype around one particular brand. The best thing you can do is go to a cooking store that lets you hold the knives; ideally, the store will let you chop some veggies with them. This was how I chose the knives that were right for me: by actually holding them in my hand to see what felt good. Any of the major brands of knives from a dedicated cooking store will use quality metals. Also, don't be afraid to mix and match brands if you like how the chef's knife feels for one brand, but you like the feel of a different brand's paring knife.

Start with those three knives, and you can slowly add more knife types based on what you like to cook. Along with your knife, purchase a honing steel to help maintain your edge. You can learn to sharpen your own knives, or the knife store you purchased from likely has a service where you can drop them off to get them sharp-ened for a small fee.

A heavy-duty stand mixer is also a versatile tool for the kitchen. Between mixing and the attachments, you can do a lot, and there is a reason why you find them in most homesteaders' kitchens. I'm personally not a huge baker, so I only have a basic KitchenAid mixer that handles most of what I need to do.

The last group of items is for all-around food preservation. Depending on what you decide to do and your interests, you'll need a few pieces of specialized equipment. In the next step, I'll review different food-preservation methods to give you an idea of what each involves.

STEP 44   PRESERVE YOUR FOOD

—

Food preservation is a hefty topic, so I'll give just a quick overview and let you take it from there. When it comes to preserving your food, you have a few approaches—each a skill in its own right.

Canning: This approach breaks down further into water-bath canning and pressure canning. The types of foods you want to can will determine the method you use. The biggest takeaway I'd like to leave you with is this: only use recipes from reputable sources, such as the USDA, the *Ball Blue Book Guide to Preserving*, or your local agricultural research university. These recipes have been tested to ensure that your canned food is safe and stays that way. Avoid using some random recipe passed down from your great-grandma or off the Internet.

The biggest safety concern in low-acid foods is botulism, which comes in three forms: the spores, the bacteria, and the toxin. The spores germinate into bacteria, which produce the toxin: the element dangerous to humans. The difficult part of botulism is that its spores are particularly heat-tolerant, only being eliminated at 240°F. The bacteria and the toxin are eliminated at boiling temperatures (above 212°F).

Because of this, some foods require you to use a pressure canner to achieve that higher temperature. The factor that determines which foods require a pressure canner is acidity; high-acid foods (a pH of 4.6 or lower) can be processed in a water-bath canner. Regardless of the method, reliable recipes should be perfectly safe, since they are developed from trusted sources and are validated in a lab, with large safety margins built in.

For both methods, you need a very large pot that can hold your cans; pressure canning requires that your pot is a pressure cooker with either a dial or weight. If you have a dial pressure canner, know that you can bring it into your state's extension office (anywhere in the United States) and they will test it for free to ensure that it's accurate.

Along with the canner, you'll need jars, lids, rings, and all your food prep equipment. You probably have most of these items already, except for the jars, lids, and rings. Remember that you need a place to store all these things. Jars and large pots can take up a lot of space, so account for that in planning your pantry.

Pickling: The main function of pickling is to preserve foods by submerging them in a high-acidic or high-salinity solution. If you think back to canning, foods that are high in acid are much easier to can because they're less susceptible to bacteria.

Pickling just takes this to an extreme, using a vinegar or salt brine solution that is so inhospitable to bacteria that it renders the food safe. You still need to follow tested recipes when you pickle, but the temperatures for processing are much lower.

Freezing: You might not have considered it, but freezing is indeed a form of food preservation. We've covered the details of chest freezers, but I wanted to share a few quick tips. When it comes to freezing, a vacuum sealer is going to be your best friend.

Removing the extra air from your packaging reduces storage space requirements, but also, by reducing voids, your food is less prone to freezer burn or ice crystals. This lets you increase dramatically the length of time you can store your meats.

Getting a vacuum sealer was a game changer for me, given its many uses. The main advantage was I could buy in bulk or whole cuts of meat. Then I'd break down the cuts myself and portion them to a size that was good for one meal. Instead of buying fillet mignon, I could buy a whole tenderloin for several dollars cheaper and cut it myself. Also, I could buy larger packs of ground beef and portion them out as I liked. After sealing, I used a rolling pin to make them flat and easy to store.

I do want to note one thing about propane-powered fridges and freezers. Several friends have gone this route and every single one of them regretted the decision. In the end, they ended up getting rid of the propane fridge/freezer and changing over to a traditional freezer running on solar-generated electricity. After making the switch, they were very happy with the setup.

Dehydrating: Next to canning and freezing, dehydrating is the most common other form of preserving foods on a homestead. Dehydrating veggies, fruit leathers, spices, and making jerky are all great options when you have a dehydrator.

The downside to these appliances is that they use a moderate amount of power for a long time, which can be challenging. The fully off-grid solution is to build your own solar dehydrator. Many of the solar ovens available for purchase have a way to dehydrate easily without any other equipment. Otherwise, I've seen many clever ways people build solar dehydrators that use only a little power to run a small fan.

Cold smoking: This is uncharted territory for me, but others who do take it on start by using a salt cure. Smoking alone doesn't render meat safe, but in combination with another preservation method, it can dry the meat thoroughly enough to make it inhospitable to bacteria. Slicing the meat thinly and salting draws out the moisture, so bacteria cannot grow. Additionally, salt is naturally antibacterial in its own right, capable of drawing moisture from bacteria, leading to cell death.

After salting the meat and draining the liquid off, you can set up a smokehouse where you carefully add smoke at a low temperature, typically under 100°F. The point here isn't to cook the meat (in fact, that is undesirable), but to slowly dry the meat even further while preventing contamination from bugs and bacteria.

Building a smokehouse lets you exclude bugs from entering, and any that do are kept at bay by the smoke. The smoke itself contains two classes of chemicals that are naturally antibacterial: phenols and carbonyls. Phenols rupture the cell walls of bacteria, killing them. Carbonyls bind with enzymes in bacteria, disrupting their function, and then bind to amino acids, which starves bacteria of a food source.

Fermentation: This breaks down into three different types: alcoholic fermentation, lactic fermentation, and acidic fermentation. Each utilizes a different chemical mechanism to create an environment that is inhospitable to bacteria.

Alcoholic fermentation converts sugars into ethanol, which is naturally acidic. This will be your wines, meads, beers, and ciders. Properly stored, you can keep these for years if necessary and still be safe.

In lactic fermentation, you leverage certain microbes to produce carbon dioxide and lactic and acetic acids, which quickly lower the pH of the environment. This kills bacteria, and by fully penetrating the vegetables, you create a food that can store for a few months up to a year. The process is similar to pickling but will produce a slightly different taste.

Finally, acidic fermentation is essentially creating vinegar. The funny thing about vinegar is that it's essentially wine that has "gone bad." So the first step is to make wine from whatever fruit you will use for your vinegar. After you create the wine, you let it continue to process all the sugars remaining until the solution turns acidic.

After the solution stops making carbonation bubbles, you'll let it continue to acidify over time, typically 20–30 days. During this time, you are essentially letting it "go bad" in a controlled manner. This sometimes means that you get a little bit of white mold that forms on the fruit pulp on top. You just scrape it off and let it do its thing.

After about 30 days, your vinegar will fully mature, and you can strain off the vinegar and filter it through a coffee filter to remove any solids. Some people like to preserve this, calling it "the mother" for its health benefits.

Cheese making: This is an art form in its own right, but it functions according to principles very similar to the rest of these methods. Essentially, you're removing moisture through curdling. Then you dry the cheese slowly over time while the cheese itself has a salty and acidic internal environment.

There are many different types of milks and techniques to make different regional cheeses, but they all function from this core principle of preservation. Reduce the moisture content and inhibit bacterial growth through an acid or salt. As with the other methods, the goal is to produce a tasty food that you can store for an extended time.

STEP 45   HEAT YOUR HOMESTEAD

—

Nothing says a cozy cabin like a woodstove. To sit in a comfy chair, sipping a cup of coffee on a cool morning while the fire crackles is a dream for many homesteaders. A woodstove is the center of many off-grid homes for a good reason; it is a great way to heat your home in a self-sufficient manner.

Modern woodstoves have come a long way, both in terms of safety and efficiency. Today's stoves burn less wood and heat your home better than ever before, but they still come with pros and cons.

Woodstoves are great at what they do: burning wood and putting out heat. They can be relied on as long as you have enough wood chopped and seasoned. That's great when you live off the grid, because heating a space is very energy-intensive and you would otherwise need a much larger solar array and battery bank.

Woodstoves can also be used for other things like drying wet clothes from the laundry, cooking, heating water, and drying herbs. There is also just the ambiance of a wood fire; the heat itself has its own quality that you don't get from a forced-air furnace.

All these advantages are balanced with the downsides, of course, namely the need to chop wood, stack wood, haul wood, and more. I grew up in a home that was heated with a woodstove, as were many of the homes in my area. I remember sometimes having to split a pile of wood before I was allowed to go play. Today, many labor-saving devices like log splitters make this task easier, but not easy.

Woodstoves also produce wood ash and wood smoke. Despite your best efforts, your home will smell like smoke from time to time, and ashes in your house will be a reality. I love the smell of a wood fire, but not in my home. Once your stove is heated up, the chimney will draw the smoke up and out of your stove, but while the stove is still cold, smoke may come into your house.

The other main downside I've experienced is that it can be hard to distribute the heat evenly throughout your house. This isn't a big issue if you have an open-concept home or a one-room cabin, but if you have several rooms, or your home has some corners that are tucked away, it can be hard to get them warm.

Just adding more wood to the fire doesn't always solve this issue, because the room with the stove becomes too hot, while the warmth doesn't flow to other areas. To combat this, you need to carefully plan the placement of your stove and then consider how you might move cold and hot air around with fans or even some ventilation ducts to heat the space evenly.

Even with good planning, unless you have a forced-air setup that distributes the heated air from the stove, your home will inevitably have cold spots, and your main room will be a bit too warm at times. This inconsistency can be a challenge in making your home comfortable and is one more thing to manage.

The advice I'll pass along is that it is far more effective to move the cold air towards the room with the stove rather than move the hot air out. Using fans to push cold air out of your colder rooms creates a thermal convection, with warm air flowing to the cold areas and the cold air flowing to the stove. Once you get this rolling with fans, it almost takes care of itself.

Remember that hot air rises, while colder air sinks. To avoid having fans all over your floors, you can mount small door fans in the top corner of your door to help push air between rooms. A better solution would be to plan for air distribution from the get-go and have a duct and fan system, but this adds a layer of complexity that many aren't willing to deal with.

The other thing about woodstoves is you must tend them, which doesn't take long once you have a bit of practice, but it isn't a matter of set-it-and-forget-it. This isn't as much of a challenge during the day, but at night, when you're sleeping, your stove will get cold as it burns down the logs you loaded right before bed.

If you've ever lived in a home with a woodstove, you know the dance you do on a very cold morning, when you slip out of your warm bed, quickly try to get the fire going again, and then run right back to bed because your house is cold from the fire dying out. It's my least favorite part of the whole affair.

Now, one solution to consider is a pellet stove. These are stoves that use small wood pellets that look like rabbit food. What is great is that they automatically feed these pellets slowly over the course of a few days at a rate that keeps your house to a certain temperature.

You not only avoid the need to constantly tend the fire or wake up to cold mornings, but your home stays roughly the same temperature over time. The two main downsides are the pellets cost about 20 percent more than split wood, and the stove itself does require power.

A pellet stove will use about 500 watts while it's lighting itself, taking about five minutes to turn on at that power. After it lights the stove, it will then continuously run a blower motor, running at about 50 watts, to push warm air into your space. While it does draw power, these levels are very easily achieved while on solar; this uses much less power than your fridge would.

The other thing I like about pellets is that you can get larger hoppers or even silos that automatically feed in the pellets. In most woodstoves, you can get through 24 hours when it's cold outside, assuming that you have a properly sized stove and that your home is reasonably insulated. If it's just slightly cool out, you can get several days out of a single hopper-load.

Larger hoppers can mean that you don't have to refill for days or even weeks. I know one guy who heats his farm shop with pellets; outside, he has a small silo that holds several tons of pellets. This feeds with an augur directly into his stove's hopper so he never has to fill it. The delivery truck puts the pellets into the silo at the start of the season, and it automatically feeds the stove all winter long.

Another option I like is the wood boiler, a furnace that is placed outside your home that heats water to around 190°F and then circulates that underground and into your home, where it can be used to warm air in a traditional HVAC unit, serve as underfloor heating through pipes in your floor or foundation, and even help heat your hot water tank.

My favorite aspect of these units is that the boiler sits outside, where you can load it without having to bring any of the mess, ashes, or wood smoke inside the home. You can set these boilers as close as 50 feet and as far as a few hundred feet from your house, which allows you to place them in an area that is easy to access but isn't an eyesore.

The other advantage of placing your fire outside the home is safety. In the worst-case scenario, your boiler could burn (though they are made mostly of metal), but it would be far away from your house. While boilers tend to use a bit more wood, people who have these rave about them.

Although boilers have their downsides, as does any heating option, they have a lot going for them, too. As mentioned above, these units can work with a traditional HVAC system to provide most of your heating; a heat exchanger transfers heat from the hot water to the moving air. Most people I know who have boilers have this setup in addition to a traditional heater so that if they can't get outside in time to refuel, the traditional heater will take over. Inside your home, you'll have a thermostat that controls the boiler's internal fan and the water pump; this means you can more precisely heat your home without being too hot or too cold. Since the boiler can use your traditional HVAC system, your air ducts will carry that heat to all parts of your home to warm it evenly.

When it comes to your water heater, you again have a heat exchanger block that warms the water most of the way, allowing you to use very little power if any to heat it to your set temperature. Even in the summer months, you can load wood into the boiler outside and it will just slow down the fan to burn more slowly. Some of my friends load their furnace once a week to have unlimited hot water with no power bill.

Another way you can use the hot water is to run the water through PEX lines embedded in your concrete slabs. This can be used to heat the whole house, depending on the setup of your home. Running the heat into your garage or workshops means when that when you're working on your vehicles and equipment in the cold winter, your floors will be warm when you lie down to get under your car.

There is also the possibility of running hot water lines under the soil or in the slab of your greenhouses. For those who have shorter growing seasons, this can double the yields by warming the ground and therefore extending your season. Just as in a slab, PEX lines are run in the ground to warm up the soil.

To keep the heat where you want it, you can dig slightly deeper than your bed typically goes, lay down a layer of closed-cell foam board, staple your lines to that, cover it with a thick piece of plastic sheeting, and then recover it with dirt. The ability to warm the dirt just enough to keep it from freezing will keep help your greenhouse weather any frosts that come through.

Alternatively, you can run the hot water through a heat exchanger in front of a fan that will heat the air of any space you want. Some space heaters are designed to be piped into hot water loops from a boiler, too.

With a combination of these methods, you can keep your home comfortable even when the cold sets in.

STEP 46   COOL YOUR HOMESTEAD

—

Cooling can be a real challenge when you're living off the grid. You simply don't have as many options to generate cold air as you do to burn fuel for hot air. The first thing to realize is that cold air isn't cold; it's just air that has less heat in it. In this way, air-conditioning really is about extracting heat from your air.

If you live in a moderate climate where the humidity is low, cooling can be a simple matter of opening your windows or using a ceiling fan. Easy tricks, including shading your home with trees, adding an awning on your windows, or utilizing passive cooling techniques to let hot air escape out the top of your home can be effective.

If, on the other hand, you live in a hot or humid area, chances are that you'll need to go beyond those basic strategies and develop a solid plan for cooling your home. No amount of open windows or passive cooling can overcome a humid heat. Nevertheless, bringing down the humidity can make high temperatures a lot more bearable.

Humidity is just moisture in the air, and water transmits heat 25 times more efficiently than dry air. So by controlling for humidity we can reduce the amount of cooling we need to do in the first place. Typically, you want your humidity to sit between 40 and 50 percent. If it's lower than that, the air feels dry and your skin will feel dry, too.

Above 50 percent, you start to feel uncomfortable, and if the humidity stays above 60 percent for extended periods, you may start to have mold issues in your home. Now, when you cool the air, there will be some dehumidification effects because as the warm, moist air runs across the cooling coils of your air-conditioner, water will condense on them just like it does on a cold glass on a hot day.

Air-conditioning is very demanding on solar systems, but a dehumidifier can be much more practical. Once you get your humidity down to a reasonable level, the unit can run intermittently to maintain it. While an AC unit can dehumidify, its best to treat these two purposes separately, allowing you to save power without having to run your AC just to manage your humidity.

Beyond managing the humidity, you might also need to cool your space further. There are three strategies that can work to combat the heat when it comes to living off the grid. The easiest one is to use fans and passive cooling in your home for as long as you can. That seems simple, but it is still worth calling out. This option is by far the least power-intensive option, even if you have to run your dehumidifier for a while to take the edge off the heat.

The next is using solar to run the most efficient mini-split that you can find. The blessing of summer is that we have a lot of sun, which means we can use that abundance to capture the sun's energy in our solar panels to power our AC.

The downside to heat is that it reduces the efficiency of your panels when they warm up. That's why placing them on the ground and not on a hot roof is a preferred option if you can. You'll want to design your racking system to allow airflow to the back of the panel so it stays as cool as possible.

Using a mini-split is key because of its terrific efficiency. Regardless of the quality of your system, you'll need a large solar array and a very large battery bank to go with it. The real challenge of cooling with solar panels comes during nighttime; in my area, the temperature drops in the evening, but it can still be in the high 80s and sometimes even in the 90s, not to mention that the humidity rises at night.

The first step in getting through a night with air-conditioning is to have a large battery bank that you can top up each day. The hardest part is when you get a week of overcast skies that block the sun, and yet it's still hot out. That's where you'll have to leverage your generator to bridge any gaps that come up.

I designed my system to be able to run for three days off the batteries with minimal sun. Except for a handful of days, that covered my needs the entire decade I lived exclusively on solar. An auto-start generator would have made the off-grid life so much more comfortable, which is why I recommend it to anyone who will listen.

One strategy I used was to "super cool" my house near the end of the day, when my panels were still producing a lot of power, but my batteries were topped off. Remember that when it comes to heating or cooling your home, the house and everything in it act as a thermal mass that holds heat. In the winter, it can take a while to warm up all the surfaces and objects in your home, but then it keeps the temperature even.

For cooling, your floors, walls, furniture, and everything else holds heat within its mass. This means that even as you cool the air, heat moves from places with higher "density" of heat to lower heat. (I'm sure physicists would have a field day with that statement, but that's how I think of it.) Effectively, this means that you need to cool your house so much that you also cool everything in it. Then, overnight, when your AC is running less, all your things aren't heating up the air as much as they would otherwise. In fact, the cool objects will slowly absorb some of the heat, drawing it out of the air. This won't mean that you can shut off your AC at night, but it will help you draw down your battery bank less.

My system used about 800 watts during the day time when running, then at night it would typically go down to about 400–500 watts. The challenge is that this draw is pretty consistent; the system doesn't run all the time, but it runs quite a bit when on really hot days and nights.

Having good insulation in your home is an important part of this; during the winter, it keeps the heat in, and during the summer, it works to keep the heat out. Remember that insulation is just a material that slows the transmission of heat from a warm place to a cool place. The respective locations of that warm place and cool place just flip-flop during the summer, so you are trying to slow heat as it seeps in.

Another option you have is geothermal cooling. This is likely not going to replace your AC unit (or your heater in the winter), but it can greatly reduce your AC use. I mentioned earlier that everything that has mass in your home needs to be cooled or heated because everything holds heat; in a way, it acts like a battery, but instead of holding electricity, the mass in your home holds heat or is cooler, so it will "recharge" like a battery by absorbing excess heat.

Taking that to an extreme, we can use the Earth itself as a giant battery for our purposes. At around thirty feet deep, the Earth's soils stay a relatively stable temperature of 55°F year-round. During the summers, we can cycle water through a closed-loop system to take the warm water and cool it in the cold of the Earth's soils. We then bring that cold water into a heat exchanger, where it cools warm air in an air handler, giving us cold air.

By cycling the water through this heat exchanger, we can slowly cool our space; the challenge I've found is that while the concept seems great on paper, the pumps to cycle the water have to be quite large and run for long enough that the power consumption really adds up.

There is also the added cost of drilling the wells to drop your cooling loops into, which can be thousands of dollars per hole, and you usually need several holes to get enough surface area to cool things. If you have a very deep lake, it's possible to run loops down to the shoreline, along the bottom of the lakebed, down to the deepest parts where it is the coldest. I've seen this in action, but it's rare.

This process can offset the cost of drilling if you have that opportunity, but it often requires environmental reviews. Even if you're able to get around the cost of drilling, by using a lake or leveraging horizontal geothermal systems, you still can't get away from the power consumption of the circulation pump. In the end, this is an available strategy, but mini-splits seem to have an edge over this approach.

STEP 47   GROW YOUR OWN FOOD

—

We've talked about considerations for growing your own food throughout this book; after all, growing your own food is a key element to becoming self-sufficient. Nevertheless, in this day and age it is extremely difficult to grow *all* your own food. While it is possible, certain areas or pieces of land tend to lend themselves to certain crops or the benefits of specializing in a crop. This means you can sell or trade your crop with others, which is its own form of self-sufficiency.

It is important to note that you don't have to raise *any* of your own food, or you can push to do it all yourself. Thinking back to your initial ideas about designing a life and homestead that is right for you, figure out what you want to do. Don't get caught up in "should" or what someone else's definition of off-grid living is.

As every farmer knows, there are seasons not only in the year but in life. You can adjust how much you grow over time to suit your changing circumstances and priorities. You might decide to start a business, go back to school, spend more time with family, or travel a bit; all those things might mean that you grow less of your food and this is fine, as long as you're intentional with how you set up your own life.

There are many great resources for information on gardening, animal husbandry, and other means of food production, so rather than trying to summarize them here, I suggest you seek them out. What I can do is share a few things that really helped me in my journey.

My first piece of advice is to start now and know that you're going to fail to some extent. A good gardener is just someone who's failed more at growing plants than you have. Gardens teach patience and instill a strong work ethic—and sometimes, despite your best efforts, they still fall flat. When it comes down to it, growing your own food is an exercise in optimism. I always encourage people to start small and grow slowly with their gardens. Learn how to manage your crops at a small scale so that when you learn hard lessons, it's easier to recover.

Over the years I've also come to appreciate smaller gardens that I grow more intensively. This approach is useful for those who don't have a lot of space, but even though I have plenty of space, I find I prefer to work in a smaller area. Growing in a smaller area is just easier to keep up with. You can control the soils better, have less ground to weed, and don't have to rely on chemicals as much as I might need to in a larger garden.

In the past, I've planted large in-ground gardens where I used tractors and tillers to make things work, but I ended up shifting back to all raised beds. The soil in my area is full of red clay, and it's hard to incorporate enough organic matter to get the soil to drain properly. No matter how many loads of mulches and amendments I added, the soil just seemed to swallow them up like it never happened.

Raised beds let me go from zero to fantastic soils in a matter of hours. I prefer taller beds, so I don't have to bend as much. In order to save on all the extra soil required, I first fill the beds with whole logs for the first 12 inches and then fill in and top with 8-plus inches of the good garden mix. There is the up-front cost of building the beds and purchasing compost, but using raised beds also means that I don't need an expensive tractor or tiller.

To compensate for the smaller space when compared to a large in-ground field, I leverage succession planting and intercropping. The first technique is just timing your plantings so that your harvest is staggered, allowing you to harvest at the same pace you eat.

One of my favorite ways to do this is to plant different varieties of the same vegetable—for example, tomatoes. I like to plant Early Girls that will be ready by about day 55, Sun Golds that come in around day 60, Romas that come in around day 75, and then Brandywine varieties that come in around day 80 or later. This way I can plant them all in one shot and they do the rest. Alternatively, you can just plant the same thing and plant a few each week.

Intercropping is mixing various plants that grow at different speeds, planting fast-growing species in between slower-growing ones. This lets you essentially double or triple your yields within the exact same space. This approach does require some more thought and planning, but once I figured out the right combinations, it worked like a charm.

One of the biggest lessons I learned that helped me grow my plot size was the importance of mulching my beds. Mulching does wonders for keeping the weeds at bay and for improving soils. I varied my mulches over the years, based on what I could get for free or cheap. Using different mulches also meant that I added different nutrient profiles as the mulch broke down.

The two things that I steer away from are hay and wood chips, though I've used both in my gardens. Hay tends to contain seeds, particularly weed seeds; instead, I've shifted to straw, which is a by-product of cereal crops like wheat. The difference is that most of the seeds in straw have already been removed, since they're what the farmer is after to begin with.

Wood chips can work well, too. I just find that most of the chips tend to be too chunky to break down fast enough. This is because most free wood chips are broken down from a chipper and not fully "mulched" into smaller pieces. I tend to like mulches that will break down most of the way in about a year, so if I decide to change it up next year, I don't have to move the mulch because it's just part of my soil now.

I also found that mulching really deep meant fewer and fewer weeds; it also builds my soils faster. I typically spread the mulch at least four inches thick, incorporating a few types in layers. I usually put lighter mulches that are more likely to blow away on the bottom, then layer heavier mulches on top. Throughout the season, I'll add around two more layers to prevent weeds from coming back.

Those are a just few personal tips for growing your own food. Take them for what you will and then learn from all the great resources that are already available.

STEP 48 BUILD A BUSINESS

—

Adding income-producing activities is a great way to offset the costs of your homestead or reduce the need to work outside your land. Of course, there is a lot to building a business, but when it comes to living off the grid, nothing makes you more self-sufficient than not relying on a paycheck from some corporate entity.

Your first step is to decide whether you want to earn money from your homestead itself. You can run your own business, but it does not necessarily need to involve farming. There is also merit in keeping your homestead separate from your income—something that you simply enjoy instead of it becoming a job; for some, that would just ruin the experience. For others, it is the only way to earn income because moving to a more rural location means fewer employment options.

I know a bit about this topic since I've started, run, and sold several businesses. My first word of caution is that you can't expect to start a business and make it your sole source of income overnight. It is rare for someone to spin up a business and earn a full livelihood within a short period of time.

Instead, when people ask for guidance on starting a business, I encourage them to begin slowly, with a clear plan, working to offset their traditional income source over time. There will come a point when you are so busy that something has to give—fortunately, that time usually coincides with when you're making enough to leave your first job.

When you arrive at this moment, try to stick it out for at least six months to ensure that the income has become stable. This allows you to earn a double income: half that money can be saved for a rainy day fund and it gives you a financial runway once you do take the leap.

This strategy relieves a lot of pressure; the bills will keep coming, and you don't want to be anxious about making the next mortgage payment. When you're less stressed, you're also more creative and open to new ideas. This is critical in starting a business because it helps you adapt and recognize opportunities.

I've seen too many people who are stressed financially fail—not because their idea wasn't good but because their survival mode hampered their ability to execute and adapt. So do your future self a favor and take some of the pressure off by being patient and building your business sustainably.

People who want to start a business also have a propensity to fall into the old "shiny object syndrome," which we've discussed in the context of homesteading projects. But it also happens with entrepreneurship: you get excited about a new idea, so you drop what you were doing and move on to the next thing, only to drop everything again when a yet another new idea comes along. The process repeats over and over, with the result that nothing ever gets brought over the finish line, but instead, you have a long trail of half-finished business ideas.

I'm not sure why so many people fall into this trap. What I do know is that many of the most successful businesses I've witnessed are pretty boring. But their owners execute well and finish what they've started efficiently.

In the farming space, one common practical business I appreciate is a CSA—community supported agriculture. The concept is simple: you find customers who want some sort of farm product, they commit to it up front financially, and then you deliver the promised goods or services.

You can sell almost anything you make on a farm, but vegetables and meats are the two most common offerings. The best part is that your costs are covered by your customer's up-front payment, which takes a lot of risk out of the endeavor. You do want to build in some margin in case something goes wrong, and this business requires a lot of planning to ensure that you have enough product to deliver each week, week after week. But unlike a farmers market model, in which your product goes to waste if you don't sell it all, you only need to invest in as much seed, soil, material, and equipment as you know you'll need.

Beyond the planning that goes into running a CSA, you need to build trust with customers before they are willing to front all that money on a promise. The larger your CSA gets, the easier this task will become, but it may take some time and persistence to establish your first few customers. Many resources about starting CSAs are available, including those that are free online or through your local extension agency.

Whether you're beginning a CSA or another endeavor, running a business is complex and rarely easy. But if you start small and plan carefully, you can build an income to support yourself and your off-grid homestead.

STEP 49 AGE IN PLACE

—

For all its many benefits, living off the grid is hard work, and many homesteaders worry about "getting too old for this" or just generally slowing down. When you're choosing this life, setting up your homestead, and building your home, think about today but also think about the future. You won't be a spring chicken forever, so plan for the later years where you might not be as mobile or spry. How do you want your life to look decades from now? Maybe more importantly, how does your life *need* to look then?

One key consideration is home design. If you have the opportunity to design your house, consider choices that will make it easier to age in place. If you're purchasing land with a house already built, you might not be able to incorporate every ideal feature, but you can still plan ahead with some remodeling. You might also add basic items to your wish list when you're house hunting, particularly if you are older.

The biggest priority is a single-level home, or at the very least, a master bedroom and other key rooms on the main floor. Being farm strong will help keep you in good health, but there will come a day where not having to climb stairs will be a great relief. Similarly, a front entry and garage that are close to the ground will allow you to install ramps should you need to use a wheelchair or scooter. Along with that, wider doors and hallways, along with an open floor plan, can make things easier if you need to use a wheelchair at some point.

Consider how you are going to power and heat your home when you are older, too. Splitting wood, stacking wood, and bringing in wood for the fire is tough work, so you might consider a heating option that is more practical as you age. In your later years, you might decide to buy firewood instead of chopping it yourself, or to install a propane tank to fuel a furnace, or to add a mini-split that can run on grid power if needed. These are relatively simple changes, assuming that you thought through them ahead of time to make sure they're possible.

For food production, consider tall raised beds with a wide lip so you can sit down rather than bending over your crop. Keeping the garden bed at 30 inches tall will take more soil, but it also may allow you to continue gardening for many more years.

Financial planning is important for everyone, on or off grid, so don't neglect this when considering your overall budget. Homesteaders tend to be frugal, but it's still necessary to set aside money for the future, particularly if you want to stay on your land rather than move into a retirement community or other living situation. Ideally, you will be able to afford to have the care you need come to you. Paying off your house, with no debt and low bills is great, but you'll also need to carry yourself through all the medical bills.

Traditional homesteaders used to address this issue by raising a lot of children and living on multigenerational farms, but much has changed since those days. Your kids might have different aspirations than you do. While they might boomerang back around to an off-grid life, just as you decided what was best for you, they should be afforded those choices that you had.

It is great if your children are willing to take care of you in your old age, weed the garden, tend the hearth, and do the chores, but your plan shouldn't rely on them to do so. Instead, set aside funds for conveniences such as firewood delivery, buying more of your food than growing it, or paying the bills when your farm income tapers off.

You might not have children, so think about how you might want to pass on your legacy. Keep in mind that not everyone wants to live your life; it was right for you, but it might not be the path that others would choose. People might desire a similar life but approach it in their own way. Handing off to the next generation is something best approached with wisdom that comes from age, managing expectations of yourself, and making plans before they are actually needed.

You also can consider hiring help. If you set aside money in advance, you will have the option to hire a farmhand who can perform chores around your homestead, a cleaning service, someone to take you to appointments, or in-home care.

To offset some of these costs, you might consider bartering for help instead of having to pay for it. For instance, I've considered setting up a small apartment above my garage to rent out. When I'm younger, the extra income can help pad out retirement funds, pay for larger purchases, etc. Later on, offering free or reduced rent in exchange for help around the farm, keeping up with the maintenance, mowing the lawns, or whatever needs to be done would be a huge asset in my older age. In short, don't get overwhelmed with worry that you need to have piles of cash on hand, but do think critically about what value you can bring to the table.

On the matter of money, I see many off-gridders also going off-grid with their cash, that is, squirreling it away "under the mattress." I knew an old timer who buried cash in mason jars in his backyard and would find them with a metal detector. The challenge with such an approach is twofold: first, it is prone to theft and second, there's no way to account for inflation.

Despite the downsides of having your money in a bank—essentially just a digital number on a spreadsheet in the cloud—it does also afford you some safety. Homesteaders like to pay for things with cash, which means they save up for purchases to stay out of debt. The challenge is that while you're saving up and holding all that money (in a bank or under your mattress), it's slowly losing its value through inflation.

This isn't a big deal if you save up for a few months for a new washer and dryer, but you can take a big hit if you save up for years on a down payment for a new home or buying land. There are ways to offset this loss to inflation—high-yield savings accounts, investing the money in the stock market or bonds or precious metals, or using the funds to expand a profitable business.

All these choices carry varying levels of risk; some require the money to be tied up for a while, whether through the rules of the investment or just in order to turn a profit after fees and taxes, but there is no single right answer. The main advice I can give you is to be aware that large sums of money will lose value over time, and you're likely better off diversifying, meaning: don't put all your eggs in one basket.

STEP 50   LIVE YOUR DREAM

—

After 49 steps, it is probably clear that a lot goes into living off the grid, and it isn't always easy. There are many benefits, but in the end these choices come down to crafting a self-sufficient life that is right for you.

What self-sufficiency means in modern times can get complicated, forcing you to consider what is possible and what is practical. The only thing I really know is that building your slice of heaven on dirt that you own is a blessing, one that we are fortunate to achieve. It's a worthy dream to work toward, and also something worth starting right where you are now.

Developing a plan to live this dream is important in order to make sure that someday doesn't turn into never. There is much to consider, and with so many moving parts, it is easy to get overwhelmed. But giving yourself some grace and being armed with a plan will make it all possible.

Becoming self-sufficient means you're taking back control over your life and building a slower existence that brings so many rewards. It is impossible to give you all the information you need to prepare for every circumstance, but my goal is to give you a starting point, steer you in the right direction, and help you learn from my mistakes. You, too, will make many, many mistakes along your journey. I hope you have the chance to pass your hard-won wisdom along to the next generation.

ABOUT THE AUTHOR

—

Living on eleven acres in a homestead powered by solar, Ryan Mitchell has helped countless people simplify their lives and go off-grid. As the best-selling author of *Tiny House Living: Ideas for Building and Living Well in Less than 400 Square Feet*, he's been featured in the *New York Times*, BBC, Associated Press, *Forbes*, *Entrepreneur Magazine*, *Mother Earth News*, *Treehugger*, and National Public Radio.

www.ingramcontent.com/pod-product-compliance
Ingram Content Group UK Ltd.
Pitfield, Milton Keynes, MK11 3LW, UK
UKHW021631190825

462027UK00023B/204